VB.Net
Programming by Example

Sun Position

High Accuracy
Solar Position
Algorithms

**a resource for programmers
& solar energy engineers**

by
John Clark Craig

Proudly Published in the USA by

Thornton Publishing, Inc
dba Books To Believe In
17011 Lincoln Ave. #408
Parker, CO 80134

Phone: (303)-794-8888

getting-published.com

Sun-Track.com

BooksToBelieveIn.com
publisher@bookstobelievein.com

ISBN: 1500179477

Dedication

This book is dedicated to all the solar energy people, past, present, and future, who are striving to help make this planet a better place. Let the light shine on with rays of hope!

Acknowledgements

There are many people who helped, in one way or the other, to bring this book into existence. My solar energy boss, good friend, and now my father-in-law, Floyd Blake, is foremost to be thanked for giving me a start in this field of research.

I'm also very grateful for EJ, Floyd's daughter and now my wife, publisher, and partner in all ways. We make a good team, again in all ways.

Jean Meeus wrote what I consider to be the best possible resource for all astronomical algorithms, titled (appropriately enough), *Astronomical Algorithms*. His valuable insight, explanations, and examples were of immense value in sifting through all the algorithms in the literature.

The people at the National Renewable Energy Laboratory, and the authors of the VSOP87 theory provided very useful resources and tools for meeting these challenges as well.

Table of Contents

Part 1

Part 2

Appendices

Part I

What's in This Book

Today, more than ever, solar energy is proving to be a viable, safe, abundant source of energy to help meet today's energy requirements around the world. At the core of just about all solar energy research, whether for site planning, or real time aiming of the most sophisticated concentrating receivers, heliostats, and photovoltaic tracking systems is the need to know exactly where the sun is, at any given time and at any given location on the earth. This book meets that need better than just about any resource available.

The author, John Clark Craig, programmed all the field control and data acquisition for several of the world's largest solar energy projects in the 1980's. Projects included the square-mile two-axis concentrating photovoltaic trackers at Carrissa Plains, California, the large Hesperia, California field of flat two-axis photovoltaic trackers, an enhanced oil recovery project using a field of heliostats and a central thermal receiver tower near Taft, California, and a variety of research and development projects at Sandia Labs, Solar 1 at Barstow, the Weizmann Institute, Tennessee Valley Authority, and elsewhere around the world. While developing software systems for these projects, John was constantly looking for the best algorithms and improved code for determining accurate sun position. Most of the available documentation was not easy to use, with bits and pieces of algorithms here and there, mixed in with poorly explained terminology, and with very little guidance for translating to programming languages he was required to use. This book is the book he wishes he had when he really needed it!

In this book, two algorithms are presented for calculating sun position. The first is of "low accuracy", determining the sun's position with a maximum error of about 0.02 degrees. Note that the sun's angular diameter in the sky is about half a degree, so this algorithm's accuracy determines the center of the sun's disk with an error of about 1/25 of its diameter. For almost all photovoltaic tracking projects this accuracy is quite sufficient, and the code is short, tight, and easy to implement or translate.

The second algorithm is much more involved, but it provides a "high accuracy" sun position that's better than just about any other algorithm you'll find, with a maximum error of about 0.00003 degrees. This is the code to use if you need the absolute best accuracy for heliostats and critically aimed solar concentrators of all types. To get to this accuracy, subtle factors such as the slight oblateness of the earth, the time it takes the sun's light to get to the earth, the angle the light arrives due to the earth's speed in its orbit, and many other fine details are taken into account.

All of the code in this book provides working example results for variables at every step of the way. This is the critical piece that makes this book a valuable resource for translating to any other programming language. As each polynomial equation or other calculation is programmed, you can check to see of the results agree with the version in this book. This is the information the author wishes he had when he first was tasked with figuring out where the sun was at.

In addition to wide-ranging Internet Google searches for information, two main sources of information and guidance must be acknowledged here. The author considers the book by Jean Meeus, titled "Astronomical Algorithms" and published by Willmann-Bell, to be the ultimate best resource for any and all astronomical algorithms. Much of the information in the two algorithms presented here came from diligent study and in-depth research of this book.

For the higher accuracy algorithm the author found the down-loadable PDF documented titled "Solar Position Algorithm for Solar Radiation Applications" to be a valuable reference as well. This publication was authored by Ibrahim Reda and Afshin Andreas and was published by the National Renewable Energy Laboratory. Both of these publications in turn used data tables and calculations from the VSOP87 theory, which you can find around the Internet in various places. The author used these resources, plus a couple of unique functions of his own design, to pull it all together into an easy-to-read, easy-to-translate source that provides exactly the information you need to be able to track the sun accurately and efficiently.

About the Author

John Clark Craig is a passionate programmer from way back. Although he's learned several other programming languages such as FORTRAN, C, C++, C#, Java, Javascript, and micro controller assembly, his heart has always been with the BASIC language. John has written almost two dozen books on a variety of programming topics, published by Microsoft Press, O'Reilly Media, and others. If you really want one of those bigger, more expensive books, check out Visual Basic 2005 Cookbook by O'Reilly Media (also available on Kindle). The code snippets in that book are still valid and very useful, mostly because the language has been very stable since the switch to the .NET Framework in February of 2002. John's work as software engineer for several of the world's largest solar energy projects during the 1980's qualifies him uniquely for authoring this book.

About this book series

This "Programming by Example" series of books on Visual Basic programming is a little different than most every book out there. Rather than being language-centric, these books are more topic-centric. That is, instead of just rehashing how to program with the latest and greatest version of Visual Basic, these books provide real-world code examples that are useful in themselves. These programs help make it easy to learn tricks and techniques of the Visual Basic programming language. Presenting new programming concepts in a relevant and experimentally-based context greatly aids in understanding and minimizing the learning curve.

Introduction to
Visual Basic Express

Programming with Visual Basic Express is easy, fun, and free. Its easy-to-read syntax makes it an ideal beginner's language, but unlike in the past, this version of the BASIC language is powerful, full-featured, and as good as it gets. Because Visual Basic Express uses exactly the same .NET Framework* libraries as all the languages in the Professional Edition of Visual Studio, including C#, it's on equal ground for speed and capabilities. The BASIC language has finally grown up and gained respectability as a programming language second to none.

*The Microsoft .NET Framework is a software framework for Microsoft Windows operating systems. It includes a large library, and it supports several programming languages which allows language interoperability (each language can utilize code written in other languages.) The .NET library is available to all the programming languages that .NET supports. ~Wikipedia

There are many good books available to teach you the full depth and breadth of this language, but these other books tend to focus on learning Visual Basic itself, unlike this "Programming by Example" series, which focuses on how to use it to program for real world tasks. The goal with this series of books is to introduce you to the language in a topic-centric way. You'll learn a lot of VB tricks along the way, but the focus will be to empower you to be able to immediately apply some great software tools for the basics of whatever topic you're interested in. Most people report they learn Visual Basic programming faster by working with examples than by digesting theory.

Installing
Visual Basic Express

The goal is to get right to the programs, procedures, and application code as soon as possible, so here you won't find long, drawn out instructions on how to install Visual Basic Express, or how to read this book - the kind of stuff that fluffs out most of the bigger, more expensive books on Visual Basic. Most readers have already gotten a version of Visual Basic up and running by the time they buy this book, but to be helpful to those who haven't gotten this far, check out Appendix A, where you'll find helpful guidance and hints on getting up and running with the free Visual Basic Express version of the language.

Where to Find
the Source Code

It's always a good idea to go through the motions of typing source code to get intimately familiar with it. However, too much typing causes decreased attention and sagging motivation. Programming should be fun, stimulating, and it should lead to creative experimentation to best master its details. Go ahead and type some of the source code listings in this book in manually, but feel free to copy-and-paste the code from the web pages provided. Here's how it works. In all cases use the following template URL to access the source code listings:

http://vb-book.com/source/ABCDEF.php

Each source code listing has this type of URL listed for easy reference. Start with the URL template above and replace the "ABCDEF" characters with the six alphabetic, upper-case letters you'll see in the code listings here in this book. Copy-and-paste* the code from the web page into your Visual Basic environment, following the guidance and using the test values presented here to verify things are working correctly and as intended.

Note that it's generally a better practice to have the Visual Basic IDE (Integrated Development Environment) create the shell of each event procedure automatically when possible, and then to copy-and-paste just the core code lines from within the procedure listings.

Part 2

Main Program

This book presents two main algorithms for calculating sun position, one for lower accuracy and the other for higher accuracy. These two algorithms are presented in the form of classes, but every effort has been made to keep the classes simple, such that it should be easy to translate the core lines of code to any other programming language of choice, whether object oriented or not.

Both classes are demonstrated in a single Visual Basic Express program comprised of a single form that provides for the entering of user data and to display the results of the calculations. Figures 1 through 3 show this main form in the development environment. Three types of controls populate this form - text boxes for all input and output data, buttons to initiate the calculations, and labels to identify all the text boxes.

Appendix B shows this form as it displays input and output data and results for the three test cases built into this main application. The user should use Figures 4 through 9 in Appendix B to verify proper operation of any program that implements the algorithms in this book.

Let's start by describing how to build this form from scratch.

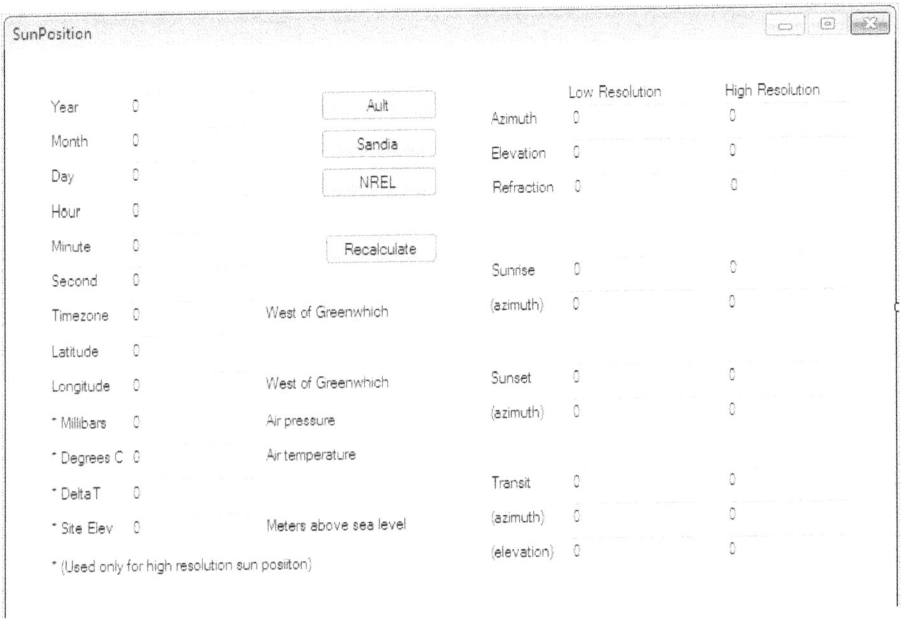

Figure 1 - The program's main form

When you first start a new Visual Basic Windows Forms Application a default Form1 is created for you. This program simply uses the Form1 all the way through. Resize the form to 700 by 450 pixels using its Size item in the Properties list. Set the form's FormBorderStyle property to FixedDialog, StartPosition to CenterScreen, MaximizeBox to False, and its Text property to "SunPosition". None of these changes are critical to the program's success, but you should get identical results to what is shown in this book by making the changes as described.

Figure 1 shows the entire form in the development environment. To make it easier to see the details for the next few steps, Figure 2 shows the left half of the form and Figure 3 shows the right half.

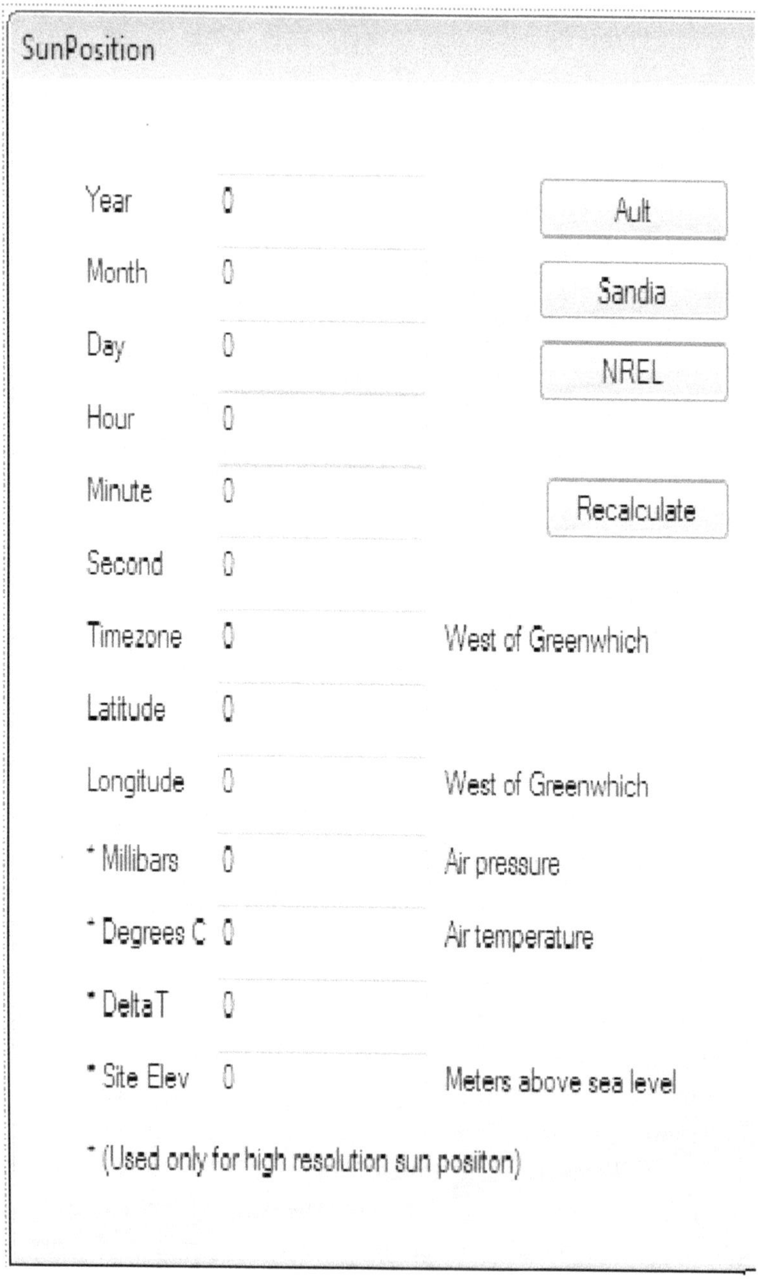

Figure 2 - The left side of the main form

On the left half of the form add a Label control and a TextBox control side by side. Change the Text property of the Label to read "Year". You can change the name of this Label control to something appropriate, but its name is not used in the programming, hence you can just let the Label control names stay at the default assigned by Visual Basic. The TextBox control is referenced in the programming to come, so change its name to txtYear. Set its Text property to "0". Size and position these two controls side by side as shown in Figure 2 for the Year label and its associated TextBox.

The next step is to repeat this process to create the whole column of Label and TextBox controls on the left side of this form. There are 13 of them, so this is a little tedious. Here's a helpful trick to simplify the process a little. Select both the Year Label control and the txtYear TextBox control at the same time. One way to do this is to click and drag to sketch a rectangle selection area that covers both controls. Press Ctrl-c to copy both controls, then press Ctrl-v to paste two copies of the originals. You'll need to drag them into the right position below the first two controls. Set the Label2 control's Text property to "Month", and rename the TextBox2 control as txtMonth. By copying and pasting in this way the new TextBox will be populated with "0" just like the first TextBox.

Repeat this process for the remaining pairs of Label and TextBox controls on the left. Refer to Figure 2 for the Text property settings for the Labels. Here's the list of TextBox control names to set:

```
txtYear
txtMonth
txtDay
txtHour
txtMinute
txtSecond
txtTimezone
txtLatitude
txtLongitude
txtMillibars
txtDegrees
txtDeltaT
txtSiteElev
```

Add 6 more label controls as shown in Figure 2. The one at the bottom displays what the asterisks in the last four TextBox controls refer to. These data items are not required for the low accuracy sun position calculations, but they are required for maximum accuracy with the high accuracy algorithm.

Place the other 5 labels to the right of the TextBoxes as shown. These are helpful reminders to let the user know the nature of the data to be entered. Some programs, for example, use negative longitudes west of Greenwich, and positive to the east. Other programs, including this one, refer to west longitudes as positive values. The one or two labels on each side of each TextBox clues the user as to the nature and units for data items to enter.

Add four buttons in the top middle of the form, as shown in Figure 2. Change the names of these buttons to btnAult, btnSandia, btnNrel, and btnCalculate respectively, and change their Text properties to read as shown in Figure 2.

The first three buttons automatically enter data for three example cases, and calls the code to calculate all results. We'll explain those routines throughout the rest of this book. "Ault" refers to a location in a farm field near Ault, Colorado, located at the latitude and longitude that was shown in the movie Close Encounters of the Third Kind. Interestingly, this location is a whole state away from Devil's Tower National Monument in Wyoming, as presented in the movie. Instead, the location at the given location of 40 degrees, 36 minutes, 10 second north and 104 degrees, 44 minutes, and 30 seconds west is in a field just outside the back window of a farm house a little ways north of Ault, Colorado. This location was chosen just for fun. To add to the fun, this example uses the date December 21, 2012, the date some people predict the end of the earth. Again, this date was chosen just for fun.

The second button presents a set of test data for a moment in a day back in 1981 when the author was working on heliostat control software at the Sandia Labs CRTF facility near Albuquerque, New Mexico. A printout of high accuracy sun positions was made available and its been used as a baseline reference by the author ever since.

The third button presents the same example data as provided in the NREL document mentioned earlier. The location of this example is at the NREL facility near Boulder, Colorado.

The fourth button is a little different from the others. If you want to alter any of the input data items, use this button to recalculate the results. Also use this button if you want to input an entirely different set of data for a location, date, and time of your own choosing. For example, to compare the results for the first example at Ault, Colorado, but for a position south of the equator instead of north, go ahead and click the Ault button, change the latitude's sign to negative, and then be sure to click the Recalculate button instead of the Ault button.

Use these three examples as test cases to verify proper results as you enter or translate the lines of code presented in this book.

All the way through this book, there are example numerical results presented for just about every intermediate variable and calculation along the way. These numbers are for the Ault, Colorado example, and they allow you to verify results at each step of your coding. This is a feature that can greatly aid in the coding or translating of these routines, and that the author wishes had been more readily available during his research to create these algorithms. Too often the sample results provided in other documentation has only appeared after multiple calculations and steps, making debugging awkward and cumbersome.

	Low Resolution	High Resolution
Azimuth	0	0
Elevation	0	0
Refraction	0	0
Sunrise	0	0
(azimuth)	0	0
Sunset	0	0
(azimuth)	0	0
Transit	0	0
(azimuth)	0	0
(elevation)	0	0

Figure 3 - The right side of the main form

On the right half of the form, there are 12 more Label controls and 20 more TextBox controls. These TextBox controls are all for output only, and hence you should set all their ReadOnly properties to True. As a result, the program sets output data by assigning strings to their Text properties, but the user can't type anything into them.

Notice that the first column of 10 TextBoxes output data results for the low accuracy calculations. The second column outputs exactly the same data items, but at the higher accuracy.

Name the low accuracy TextBox controls as follows, from top to bottom:

```
txtAzLo
txtElLo
txtRefractionLo
txtSunriseLo
txtSunriseAzLo
txtSunsetLo
txtSunsetAzLo
txtTransitLo
txtTransitAzLo
txtTransitElLo
```

Name the TextBoxes in the right column exactly the same, except replace all occurrences of "Lo" in their names to "Hi". So, for instance, the top TextBox in the rightmost column is named txtAzHi.

The form's code

You may copy-and-paste the following code block from
http://vb-book.com/source/PYTJIF.php

```
Public Class Form1

Private Sub btnAult_Click(ByVal sender As System.Object,
ByVal e As System.EventArgs) Handles btnAult.Click
    txtYear.Text = "2012"
    txtMonth.Text = "12"
    txtDay.Text = "21"
    txtHour.Text = "12"
    txtMinute.Text = "12"
    txtSecond.Text = "12"
    txtTimezone.Text = "7"
    txtLatitude.Text = "40.6027778"
    txtLongitude.Text = "104.7416667"
    txtMillibars.Text = "820"
    txtDegC.Text = "-5"
    txtDeltaT.Text = "67"
    txtSiteElev.Text = "1519"
    CalculateLowAndHigh()
End Sub

Private Sub btnSandia_Click(ByVal sender As System.Object,
ByVal e As System.EventArgs) Handles btnSandia.Click
    txtYear.Text = "1981"
    txtMonth.Text = "6"
    txtDay.Text = "21"
    txtHour.Text = "8"
    txtMinute.Text = "0"
    txtSecond.Text = "0"
    txtTimezone.Text = "7"
    txtLatitude.Text = "34.9622313889"
    txtLongitude.Text = "106.509031111"
    txtMillibars.Text = "840"
    txtDegC.Text = "19"
    txtDeltaT.Text = "51.8"
    txtSiteElev.Text = "1707"
    CalculateLowAndHigh()
End Sub

Private Sub btnNrel_Click(ByVal sender As System.Object,
ByVal e As System.EventArgs) Handles btnNrel.Click
```

```
txtYear.Text = "2003"
txtMonth.Text = "10"
txtDay.Text = "17"
txtHour.Text = "12"
txtMinute.Text = "30"
txtSecond.Text = "30"
txtTimezone.Text = "7"
txtLatitude.Text = "39.742476"
txtLongitude.Text = "105.1786"
txtMillibars.Text = "820"
txtDegC.Text = "11"
txtDeltaT.Text = "67"
txtSiteElev.Text = "1830.14"
CalculateLowAndHigh()
End Sub

Private Sub btnCalculate_Click(ByVal sender As System.Object,
ByVal e As System.EventArgs) Handles btnCalculate.Click
CalculateLowAndHigh()
End Sub
```

The operation of this form is fairly straightforward. When any of the first three buttons are clicked, data is automatically entered into the TextBox controls on the left, exactly as if the user had typed them in very quickly. A subroutine called CalculateLowAndHigh() is then called. This routine creates an instance of the SunPositionLowRes object called spLo, as defined in a class module of the same name which will be described in more detail later. This object has properties corresponding to each of the data entry items. For example, the Year property is assigned the year number from the txtYear TextBox control. Once all of the properties have been assigned, the object's SunPos() method is called to do the calcualtions. Results of the calculations are then available in properties such as Azimuth, Elevation, and so on. These values are then copied out in string form to the output TextBoxes on the right side of the form.

Similarly, once all the low accuracy results are output to the appropriate TextBoxes, a new object instance is created from the SunPositionHighRes class, called spHi, and the whole process is repeated. The only difference is that this higher accuracy object has a few more properties for input of the air temperature, barometric pressure, height of the location above sea level, and the current value of DeltaT (DeltaT is defined later). These values are used for the more exacting calculations for apparent sun position due to the non-spherical shape of the earth and for better atmospheric refraction correction.

The output properties for the high accuracy results are identical to the properties for low accuracy resluts, except that there are more digits of accuracy following the decimal points. Hence, the results are displayed in the output TextBoxes with a few more digits after the decimal points.

The first code to add to this application is for its main form. If you are very careful to name each button control exactly as above, then you can copy-and-paste the code directly into the program. However, a better and safer approach is to double click on the buttons to let Visual Basic automatically create the template code for each button's click event, then copy-and-paste, or type-in, just the lines of code from within these events. This can prevent hassles if, for instance, the name of the button varies slightly from what's shown here. The other methods in this code can be safely copied and pasted into place.

The first block of code. listed above, shows the click events for each of the four buttons. This is where all the action begins for the user. The first three buttons load the TextBoxes with data for a given example time and place, and then call the CalculateLowAndHigh() method to do the sun position calculations. The fourth button skips all the preloading of the TextBoxes and assumes the user has entered, or modified, data of their own choosing. It just calls the calculation routine right away.

The following block of code should be added immediately after the first block, still within the Form1 Class code.

You may copy-and-paste the following code block from
http://vb-book.com/source/BPCARP.php

```
Private Sub CalculateLowAndHigh()

    Dim theTime As Date

    'Low resolution sun position
    Dim spLo As New SunPositionLowRes

    'Load the data
    spLo.Year = Integer.Parse(txtYear.Text)
    spLo.Month = Integer.Parse(txtMonth.Text)
    spLo.Day = Integer.Parse(txtDay.Text)
    spLo.Hour = Integer.Parse(txtHour.Text)
    spLo.Minute = Integer.Parse(txtMinute.Text)
    spLo.Second = Integer.Parse(txtSecond.Text)
    spLo.Timezone = Double.Parse(txtTimezone.Text)
    spLo.Latitude = Double.Parse(txtLatitude.Text)
    spLo.Longitude = Double.Parse(txtLongitude.Text)

    'Sun position low resolution at the given moment
    spLo.SunPos()

    'Output sun position
    txtAzLo.Text = spLo.Azimuth.ToString("###0.00")
    txtElLo.Text = spLo.Elevation.ToString("###0.00")
    txtRefractionLo.Text = spLo.Refraction.ToString("0.00")

    'Sunrise
    theTime = spLo.Sunrise()
    txtSunriseLo.Text = theTime.ToLongTimeString
    txtSunriseAzLo.Text = spLo.Azimuth.ToString("###0.00")

    'Sunset
    theTime = spLo.Sunset()
    txtSunsetLo.Text = theTime.ToLongTimeString
    txtSunsetAzLo.Text = spLo.Azimuth.ToString("###0.00")

    'Transit
    theTime = spLo.Transit()
    txtTransitLo.Text = theTime.ToLongTimeString
    If spLo.Azimuth 359 Then spLo.Azimuth = 0
    If spLo.Azimuth 179 Then spLo.Azimuth = 180
    txtTransitAzLo.Text = spLo.Azimuth.ToString("###0")
    txtTransitElLo.Text = spLo.Elevation.ToString("###0.00")
```

This second block of code lists the first half of the CalculateLowAndHigh() method, the part that calls the routines to calculate the low accuracy sun position results. It first creates a SunPositionLowRes object named spLo, then loads all relevant data from the TextBoxes into the corresponding properties of this object, then calls the SunPos() method to calculate sun position. The results are formatted into strings showing the proper number of digits for the accuracy provided, and loads these string results into the output TextBoxes on the right half of the form.

In addition to sun position for one moment in time, there are additional methods in the spLo object to calculate the times of sunrise, sunset, and the transit of the sun for the given date. These methods are each called in turn, as shown, and results are output in the appropriate TextBoxes. All of these core calculations are described in greater detail later in this book.

The last block of code for Form1 continues the CalculateLowAndHigh() method, repeating much of the same code as before, but in this case for the high accuracy calculations. Add this code block immediately following the previous one.

You may copy-and-paste the following code block from
http://vb-book.com/source/RJRJWE.php

```
'High resolution sun position
Dim spHi As New SunPositionHighRes

'Load the data
spHi.Year = Integer.Parse(txtYear.Text)
spHi.Month = Integer.Parse(txtMonth.Text)
spHi.Day = Integer.Parse(txtDay.Text)
spHi.Hour = Integer.Parse(txtHour.Text)
spHi.Minute = Integer.Parse(txtMinute.Text)
spHi.Second = Integer.Parse(txtSecond.Text)
spHi.Timezone = Double.Parse(txtTimezone.Text)
spHi.Latitude = Double.Parse(txtLatitude.Text)
spHi.Longitude = Double.Parse(txtLongitude.Text)
spHi.Millibars = Double.Parse(txtMillibars.Text)
spHi.DegC = Double.Parse(txtDegC.Text)
```

```
        spHi.DeltaT = Double.Parse(txtDeltaT.Text)
        spHi.SiteElevation = Double.Parse(txtSiteElev.Text)

        'Sun position high resolution at the given moment
        spHi.SunPos()

        'Output sun position
        txtAzHi.Text = spHi.Azimuth.ToString("###0.00000")
        txtElHi.Text = spHi.Elevation.ToString("###0.00000")
        txtRefractionHi.Text = spHi.Refraction.ToString("0.00000")

        'Sunrise
        theTime = spHi.Sunrise()
        txtSunriseHi.Text = theTime.ToLongTimeString
        txtSunriseAzHi.Text = spHi.Azimuth.ToString("###0.00000")

        'Sunset
        theTime = spHi.Sunset()
        txtSunsetHi.Text = theTime.ToLongTimeString
        txtSunsetAzHi.Text = spHi.Azimuth.ToString("###0.00000")

        'Transit
        theTime = spHi.Transit()
        txtTransitHi.Text = theTime.ToLongTimeString
        If spHi.Azimuth 359 Then spHi.Azimuth = 0
        If spHi.Azimuth 179 Then spHi.Azimuth = 180
        txtTransitAzHi.Text = spHi.Azimuth.ToString("###0")
        txtTransitElHi.Text =
spHi.Elevation.ToString("###0.00000")

End Sub

End Class
```

This third block of code completes the code for Form1. For the high accuracy calculations an object named spHi is created from the SunPositionHighRes class, which will be described in much more detail later on. This block of code is very similar to the code for the low accuracy calculations, but a few more data items are required for the full accuracy. Each of these four data items is interesting in how they affect the apparent sun position.

The atmospheric pressure, in millibars, and the atmospheric temperature, in degrees Celsius, are used for higher accuracy atmospheric refraction adjustments for the sun's elevation. Basically, cold dense air has a higher index of refraction, so in the winter the sun appears very, very slightly higher in the sky than in the summer. It's not actually at a higher angle above the horizon, it's just that it appears that way due to the refraction of the atmosphere. This refraction adjustment is much more significant for lower elevations of the sun, all year round. In fact, an interesting tidbit of knowledge is that when the sun appears to be just barely sitting with its bottom edge on the horizon, in actuality the sun is sitting just barely but completely below the horizon. Refraction near the horizon works out to approximately the same angular amount as the angular diameter of the sun's disk. A standard refraction correction calulation that doesn't require atmospheric pressure and temperature is used for the lower accuracy calculations, but these data items are required for the higher accuracy.

DeltaT is a measure of the difference in time between Universal Time and Terrestrial Time. You can read up in detail on this subject at many places on the Internet, such as at http://en.wikipedia.org/wiki/DeltaT. Basically, the clocks we use day-to-day use Universal Time, which is a function of the spin rate of the earth on its axis. The earth's spin rate varies over time, so every few years a leap second is added to our clocks to keep them in sync with the spin of the earth. Terrestrial Time is a more uniformly smooth time base, and it is not affected or adjusted for the spin of the earth. Events out in space, such as for example, the position of Jupiter in its orbit around the sun, are not affected by the spin rate of the earth. So, to calculate astronomical events such as planetary positions, comets, eclipses, and so on, Terrestrial Time provides a much more stable and predictable time base. Unfortunately, with Terrestrial Time the leap seconds start to add up over the years, and hence these two types of clocks are drifting apart from each other. This accumulating difference in time is called DeltaT, and it's not easy to calculate what it will be in the future very accurately. There's a parabolic curve fit that comes close over the centuries, but it's far from perfect. The best estimation calculations for DeltaT published anywhere are used in the SunPositionHighRes class, but the user may enter more precise values as published on the Internet each year if desired. DeltatT is only used for the higher accuracy calculations.

The site elevation, or height above sea level of the observer, provides another slight adjustment that's important for the higher accuracy sun position but not for the lower accuracy formulas. The earth is very nearly spherical, but not quite. It bulges at the equator slightly due to centrifugal force as it spins in space. This slight oblateness of the earth changes the direction of the center of gravity of the earth for observers as a function of latitude, and hence it changes the apparent position of space objects, including the sun, ever so slightly. The difference is enough to be important for the higher accuracy calculations.

The next two chapters present and describe the low and high accuracy sun position class modules. If your application is, for example, photovoltaic panel tracking of the sun, the low accuracy class is plenty accurate enough. Even if your equipment involves sunlight concentrating lenses or mirrors, the 0.02 degree maximum error of the low accuracy calculations is likely to be well within the tolerance range or aiming accuracy of your equipment. As a rough guide, this maximum error is about 1/25 of the angular diameter of the sun's disk. For much greater accuracy, and if you don't mind using a bigger chunk of code, or if the aiming accuracy your equipment requires is important, then the 0.00003 degree maximum error of the higher accuracy class is pretty much the best you can find anywhere.

SunPositionLowRes Class - Low Accuracy Sun Position

Add a new class module to your SunPosition project, and name it SunPositionLowRes. Add each of the following blocks of code to this class, in the order presented, to complete all the code for this class. Code for each block is described as each block is presented.

You may copy-and-paste the following code block from
http://vb-book.com/source/YEVLYP.php

```
Imports System.Math

Public Class SunPositionLowRes

'Inputs
Public Property Year As Integer
Public Property Month As Integer
Public Property Day As Integer
Public Property Hour As Integer
Public Property Minute As Integer
Public Property Second As Integer
Public Property Timezone As Double
Public Property Latitude As Double
Public Property Longitude As Double

'Outputs
Public Property Azimuth As Double
Public Property Elevation As Double
Public Property Refraction As Double

Private Enum rst As Integer
   sunrise
   sunset
   transit
End Enum
```

The low accuracy sun position algorithm is presented here as a Visual Basic class, with the latest style of properties and methods using a very clean, simple programming stye. Every attempt has been made to make this code easy to translate to any other programming language of your choice. For example, the properties of this class use the simplest syntax provided in Visual Basic, with no Read Only settings for output variables, or error checking during input. Feel free to add this extra layer of coding if your application demands complete error checking, or needs to guide the user as to proper range of values for input variables, and so on.

The one-line syntax for properties is new in the latest version of Visual Basic Express. It is strongly recommended that the user upgrade to the latest version, although these properties could be rewritten using the older, more unfolded lines of code that accomplish the same result. For comparison, here's an example showing how the Year property could be rewritten using the older syntax:

```
Private _Year As Integer
Public Property Year As Integer
  Set(ByVal value As Integer)
    _Year = value
  End Set
  Get
    Return _Year
  End Get
End Property
```

The much shorter syntax supported in the latest version of Visual Basic is much easier to read, easier to translate to some other language, and provides identical results. The only catch is that the user's calling program can, for example, set the value of the Azimuth property to some numerical value, when the proper way to use this property is just for accessing one of the results of the calculations. It's a tradeoff situation, but the simpler syntax was chosen here to aid in understanding and to make translation easier.

The enumerated constants provide readable values that define which daily event is to be calculated later in the class. Much of the code for finding the time of sunrise, sunset, or transit of the sun is the same in each case, so this constant determines which calculation path to take for those parts of the code that do differ. This will become clearer in the explanation for how these events are calculated.

The main method of the SunPositionLowRes class is SunPos(). After all the input properties have been set to determine a precise date, time, and location on the earth, this method calculates the azimuth and elevation angle for the sun's apparent position in the sky, accurate to 0.02 degrees. This is the maximum error over time, and in most cases the accuracy is much better.

A description of the calculations follows, along with example numerical values for each step of the way. If the user is translating to another programming language, these numbers can help immensely with verifying the new code.

You may copy-and-paste the following code block from
http://vb-book.com/source/WGKGXO.php

```
Public Sub SunPos()

  'Constants
  Const tau As Double = PI + PI
  Const rpd As Double = PI / 180

  'Local variables
  Dim rlat, rlon, utim, dnum As Double
  Dim slon, sano, ecli, obli As Double
  Dim rasc, decl, stim, hang As Double
  Dim targ As Double

  'Convert latitude and longitude to radians
  rlat = _Latitude * rpd
  rlon = _Longitude * rpd

  'Decimal hour of the day at Greenwhich
  utim = _Hour + _Timezone + _Minute / 60 + _Second / 3600
```

```
'Days from J2000, accurate only from 1901 to 2099
dnum = 367 * _Year - 7 * (_Year + (_Month + 9) \ 12) \ 4
+ 275 * _Month \ 9 + _Day - 730531.5 + utim / 24

'Mean longitude of the sun
slon = Range(dnum * 0.01720279239 + 4.894967873, 0, tau)

'Mean anomaly of the Sun
sano = Range(dnum * 0.01720197034 + 6.240040768, 0, tau)

'Ecliptic longitude of the sun
ecli = slon + 0.03342305518 * Sin(sano) + 0.0003490658504
* Sin(2 * sano)

'Obliquity of the ecliptic
obli = 0.4090877234 - 0.000000006981317008 * dnum

'Right ascension of the sun
rasc = Atan2(Cos(obli) * Sin(ecli), Cos(ecli))

'Declination of the sun
decl = Asin(Sin(obli) * Sin(ecli))

'Local sidereal time
stim = Range(4.894961213 + 6.300388099 * dnum - rlon, 0,
tau)

'Hour angle of the sun
hang = Range(stim - rasc, 0, tau)

'Local elevation of the sun
_Elevation = Asin(Sin(decl) * Sin(rlat) + Cos(decl) *
Cos(rlat) * Cos(hang))

'Local azimuth of the sun
_Azimuth = Atan2(-Cos(decl) * Cos(rlat) * Sin(hang),
Sin(decl) - Sin(rlat) * Sin(_Elevation))

'Convert azimuth and elevation to degrees
_Azimuth = Range(_Azimuth / rpd, 0, 360)
_Elevation = Range(_Elevation / rpd, -180, 180)

'Refraction correction
targ = (_Elevation + (10.3 / (_Elevation + 5.11))) * rpd
_Refraction = (1.02 / Tan(targ)) / 60

'Adjust elevation for refraction
_Elevation += _Refraction

End Sub
```

Microsoft engineers decided to automatically create protected local variables for the properties using a leading underscore and the same name as the property. This makes it easy to identify class level variables, both the ones that provide input data for the calculations, and the ones that are set to provide output results.

Two constants are defined to aid with conversions between radians and degrees. Visual Basic, like most other languages, calculates trigonometric functions using radians. The various equations in the literature for astronomical calculations, on the other hand, are almost always in terms of degrees. It's important to use the correct angular units, and this is often the source of subtle bugs and hassles when adopting calculations for any given programming language.

For efficiency, compactness, and speed, this code maintains many of the intermediate calculation results in radians, converting back to degrees only where necessary.

One of the contants, named tau, is twice the value of pi. For some very interesting and fun side reading, check out "The Tau Manifesto" by Michael Hartl, at http://tauday.com/. It's all in fun, but there really are some very good reasons why tau is a better choice than 2 * pi for many calculations, including some here in the SunPos() method.

A few of the variables in this program, including several of the properties, are declared as Integer. In Visual Basic these are 32-bit (4-byte) integers that range in value from -2,147,483,648 through 2,147,483,647. Many of the variables are declared as Doubles, which are signed IEEE 64-bit (8-byte) double-precision floating-point numbers that range in value from -1.79769313486231570E+308 through -4.94065645841246544E-324 for negative values and from 4.94065645841246544E-324 through 1.79769313486231570E+308 for positive values. Most languages have variables of the same, or very nearly the same, type and range. Users translating this code to another language should have no problem determining equivalent types of numeric variables to use.

The following data provides a checklist of values to aid the user in verifying proper operation of this method. This can be valuable when hand coding in Visual Basic, and extremely valuable and time saving when translating to another programming language. Also provided is a nut-shell description of the astronomical terminology involved. Much of this terminology is beyond the scope of this book, but for the best understanding of how and why these calculations are defined the way they are, the user is encouraged to learn more on the Internet or in books. The author's favorite resource is Astronomical Algorithms, by Jean Meeus.

First, here are the date, time, and location properties used to create all the intermediate numerical results that follow. These properties must be set to these values to see the same results during debugging and testing.

```
Year 2012
Month 12
Day 21
Hour 12
Minute 12
Second 12
Timezone 7
Latitude 40.6027778
Longitude 104.7416667
```

The following values can be verified, as the user codes, by setting breakpoints just after each calcualtion, or by adding output statements to see and verify the intermediate results.

```
tau 6.2831853071795862
rpd 0.017453292519943295
```

As described above, these are two constants used in SunPos() to define 2 * pi and a conversion factor for conversions between radians and degrees.

```
rlat 0.70865215806565962
rlon 1.8280869479415036
```

All input and output properties assume degrees rather than radians. rlat and rlon are internal to the class, and they store the latitude and longitude in radians for efficiency in the calculations that follow.

The program that creates an instance of the SunPositionLowRes class should set latitude and longitude values using decimal degrees. If latitude and longitude are provided in degrees, minutes, and seconds then convert to decimal degrees using this formula:

$$decimalDegrees = degrees + minutes / 60 + seconds / 3600$$

```
utim 19.203333333333333
```

utim is the time at Greenwich for the moment provided by the local time, in decimal hours. The time zone offset shifts local time by the number of hours required, and the minutes and seconds are converted to fractions of an hour, and added in to provide a single numerical result.

```
dnum 4738.3001388888888
```

Julian day numbers provide a simple and consistent way to determine the number of days between any two dates. For astronomical calculations, this is the only way to go, because of all the complications with the number of days in the various months, which years are leap years, and so on. Fortunately, there are formulas that accurately convert from month, day, and year numbers to Julian date.

In this case, a modified Julian date is used, defined as starting in the year 2000. The conversion used here is short and sweet, but it's only accurate between the years 1901 and 2099. For sun positions outside this date range, the higher accuracy sun position can be used.

For users wanting to study how this same calculation works using Visual Basic's highly efficient Date and TimeSpan objects, this snippet of code provides a good starting point, and it can be used to calculate dnum if desired:

```
Dim dat As New Date(_Year, _Month, _Day)
dnum = dat.Subtract(#1/1/2000 12:00:00 PM#).TotalDays + utim /
24
```

This approach was decided against by the author in order to facilitate translation of this algorithm to other programming languages.

```
slon 4.7255524504790927
sano 6.0667302258499305
```

The movement of the earth around the sun is usually expressed, in the astromnomial algorithm literature, as though the sun were orbiting around the earth. This orbital motion is an ellipse instead of a perfect circle, otherwise it would be easy to calculate the sun's position against the background of the stars. The Mean Longitude and Mean Anomaly of the Sun have to do with the average and corrected position of the sun along this non-circular orbital path.

```
ecli 4.7182277835195876
```

The sun appears to orbit the earth annually exactly in the plane of the ecliptic, hence the term ecliptic longitude for its position against the background of the stars. The speed of the sun appears to vary a little throughout the year, due to the earth's slightly elliptical orbit, so adjustments are made to the ecliptic longitude using the mean longitude and mean anolmaly values previously calculated.

```
obli 0.40905464382465134
```

The earth's axis of rotation is tilted a little over 23 degrees from the plane of the ecliptic, and this tilt is changing over time. obli is this slowly changing tilt expressed here in radians.

```
rasc -1.5644325031162596
decl -0.40904725436667339
```

Right ascension and declination are astronomical coordinates used to pinpoint stars in space. For most stars these coordinates are basically fixed, but some stars do drift slowly in various directions, hence their right ascension and declination slowly change. The sun, however, is zooming around in this coordinate system, making a full circle once a year. The calculations presented so far provide the sun's right ascension and declination coordinates over time, again in radians.

```
stim 4.7832844004477906
```

Sidereal time clocks run slightly faster than normal clocks. A sidereal day is approximately 23 hours, 56 minutes, and 4.091 seconds long. This time base defines a "sideral day" as the time it takes for a star to appear to spin around the earth daily, whereas a normal clock defines a day as how long it takes the sun to appear to rotate around the earth once. The difference is that over a year the stars make an extra spin around the earth as compared to the sun, due to the earth's orbit around the sun. The public likes to use clocks based on daylight, whereas astronomers like to use clocks based on the motion of the stars.

The formula presented here calculates local sidereal time for the given moment in time and based on the observer's longitude on the earth. There are no time zones for sidereal time, just an adjustment for exactly how far around the earth the observer is located.

```
hang 0.064531596384464152
```

The hour angle of a star, or the sun in this case, is a function of sidereal time and the right ascension of the star or the sun. This defines where the sun is in the sky local to the observer, but only relative to the earth's spin axis. The conversion to azimuth and elevation are next in line.

```
_Elevation 0.45148488459965758
_Azimuth -3.0757898405426749
```

Standard trigonometric conversions allow converting local hour angle and declination coordinates to azimuth and elevation. The results are first expressed in radians, and in the next step will be converted to degrees for output to the calling program.

```
_Azimuth 183.77022346768825
_Elevation 25.868178401511386
```

The azimuth and elevation are converted from radians to degrees, with the results adjusted into the desired range of values by the Range function. For example, the azimuth of the sun is best expressed as an angle measured east from due north. 90 degrees is to the east, 180 degrees is south, 270 degrees is west, and so on. Elevation, however, is best expressed as an angle ranging from 90 degrees at the zenith, or straight overhead, to 0 degrees when the sun is on the horizon, and on down to negative elevation numbers when the sun is below the horizon. For this reason the elevation is adjusted into the circular range from -180 to +180 degrees.

The Range function is defined as a private method of the SunPositionLowRes class, and it will be described in more detail a little later on.

```
targ 0.45728796671202093
_Refraction 0.034547541310086173
```

One final correction is required to determine a more accurate apparent sun position. The earth's atmosphere has an index of refraction that bends light coming from the sun, in the same way that a camera lens or a prism bends light. When the sun is high overhead the refraction effect is minimal, but near the horizon the sun's light is bent approximately by the same angle as the diameter of the sun's disk. This means that when the sun appears to be just sitting on the horizon, it's actually just below it. For accurate solar position, this refraction correction must be taken into account and the apparent elevation adjusted appropriately. The calculation is presented in two steps, with the variable named targ (a shortened reference to "argument to the tangent function") calculated first. Its value is shown above, along with the refraction correction amount in degrees.

```
_Elevation 25.902725942821473
```

As a final step, the refraction correction amount is added to the calculated elevation of the sun to "lift" it a little in the sky by the predicted amount caused by atmospheric refraction.

Notice that Refraction is an output property of this class. There are times when the direction of the sun is required without any refraction correction. The user may subtract the value of Refraction from the value of Elevation to remove the refraction adjustment. Later on in this class, there are methods for calculating the officially defined time of sunrise and sunset, and the uncorrected sun elevation is required during these calculations.

Supporting Procedures

Almost all of the low accuracy calculations reside in the one method listed above. However, there are two separate routines that this code calls, as follows.

You may copy-and-paste the following code block from
http://vb-book.com/source/HJIKLP.php

```
Private Sub SetDateTime(ByVal dt As Date)
    _Year = dt.Year
    _Month = dt.Month
    _Day = dt.Day
    _Hour = dt.Hour
    _Minute = dt.Minute
    _Second = dt.Second
End Sub


Private Function Range(ByVal X As Double, ByVal Rmin As
Double, ByVal Rmax As Double) As Double
    Dim Delta As Double = Rmax - Rmin
    Return (((X - Rmin) Mod Delta) + Delta) Mod Delta + Rmin
End Function
```

The SetDateTime() procedure separates out and copies the individual date and time data stored in a single Date variable into the properties of this class. Individual properties were created for year, month, day, hour, etc., to facilitate translation to other programming languages. VB.Net, and other .Net framework languages such as C#, provide a Date variable to hold all data for a single moment in time in one variable. This is very useful for date and time calculations, but it can be problematic for easy translation to certain programming languages. The sunrise, sunset, and transit routines that follow use the date variables, with an explanation of how to approach translation if required, but the main sun position stays away from using the date variables. The SetDateTime() procedure allows the interfacing of these two different approaches to designating a moment in time.

The author created the Range() function to solve several related common programming tasks. Some of the astronomical calculations return an accumulated angle based on the date and time. Some of these angles are many multiples of a whole circle. For instance, the variable sano is the mean anomaly of the sun and it is calculated in the given example to be 86.406961443813714, an angular amount of almost 14 times around a complete circle, expressed in radians. The Range function brings this angle into the range of 0 to tau, which is the same as 0 to 360 degrees. This is a much more manageable angular value for further computations.

The Range() function is more than just a glorified MOD function, as it can use other ranges too. The elevation of the sun is best expressed as an angle between -90 and +90 degrees, where 0 is on the horizon, 90 is straight overhead at the zenith, and all negative angles are below the horizon. For this reason, the Range function is used later in this procedure to bring a calculated elevation angle into this range.

The time of sunrise and sunset has been defined, by convention, to occur when the sun's elevation is at -0.8333 degrees, without any refraction correction. Formulas have been developed to more or less directly calculate an approximate time of day, and at a given location on the earth, for sunrise and sunset. The author chose to instead use a binary search algorithm to narrow in on the exact time when the sun is at the target elevation, either early in the day or in the evening, as a more exacting way to determine the time of sunrise and sunset. The same binary search algorithm was also modified slightly to incorporate a search for the time of transit, or the moment when the sun crosses a due south (or due north depending on location and date) line at "solar noon" once each day. Early implementers of sun position calculations probably avoided using a binary search for speed reasons, but with today's much faster computers, and the fact that these events only need to be calculated once per day, the author was happy with his decision to use a binary search. The speed is excellent, and the time for these events is accurate to the nearest second for the accepted definition of sunrise and sunset.

You may copy-and-paste the following code block from
http://vb-book.com/source/JHFFOY.php

```
Public Function Sunrise() As Date
   Return (DailyEvent(rst.sunrise))
End Function

Public Function Sunset() As Date
   Return (DailyEvent(rst.sunset))
End Function

Public Function Transit() As Date
   Return (DailyEvent(rst.transit))
End Function

Private Function DailyEvent(ByVal dayevent As rst) As Date

   'Local variables
   Dim time1, time2, time3 As Date
   Dim azim1, azim2 As Double
   Dim elev1, elev2 As Double
   Dim msec As Double

   'Select bracketing times for binary search
   Try
     Select Case dayevent
       Case rst.sunrise
          time1 = New Date(_Year, _Month, _Day, 0, 0, 1)
          time2 = New Date(_Year, _Month, _Day, 12, 0, 0)
       Case rst.sunset
          time1 = New Date(_Year, _Month, _Day, 12, 0, 0)
          time2 = New Date(_Year, _Month, _Day, 23, 59, 59)
       Case rst.transit
          time1 = New Date(_Year, _Month, _Day, 6, 0, 0)
          time2 = New Date(_Year, _Month, _Day, 18, 0, 0)
     End Select
   Catch
     _Azimuth = 0
     _Elevation = 0
     Return New Date(0)
```

```
End Try

'Calculate sun position for first time
SetDateTime(time1)
Me.SunPos()
azim1 = _Azimuth
elev1 = _Elevation

'Calculate sun elevation for second time
SetDateTime(time2)
Me.SunPos()
azim2 = _Azimuth
elev2 = _Elevation

'Bail out if sun doesn't behave
Select Case dayevent
  Case rst.sunrise, rst.sunset
    If Sign(elev1) = Sign(elev2) Then
      Return New Date(0)
    End If
  Case rst.transit
    If Sign(180.0 - azim1) = Sign(180.0 - azim2) Then
      Return New Date(0)
    End If
End Select

Do
  'Average the two times
  msec = time2.Subtract(time1).TotalMilliseconds / 2
  time3 = time1.Add(TimeSpan.FromMilliseconds(msec))

  'Calulate sun position for new time
  SetDateTime(time3)
  Me.SunPos()

  'Determine which time, azimuth, and elevation to replace
  Select Case dayevent
    Case rst.sunrise
      If _Elevation - _Refraction + 0.8333 < 0.0 Then
        time1 = time3
        azim1 = _Azimuth
        elev1 = _Elevation
      Else
        time2 = time3
        azim2 = _Azimuth
        elev2 = _Elevation
      End If
    Case rst.sunset
      If _Elevation - _Refraction + 0.8333 > 0.0 Then
        time1 = time3
        azim1 = _Azimuth
        elev1 = _Elevation
      Else
        time2 = time3
        azim2 = _Azimuth
        elev2 = _Elevation
      End If
    Case rst.transit
      If Sign(180.0 - _Azimuth) = Sign(180.0 - azim1) Then
```

```
            time1 = time3
            azim1 = _Azimuth
            elev1 = _Elevation
        Else
            time2 = time3
            azim2 = _Azimuth
            elev2 = _Elevation
        End If
    End Select

    'Number of milliseconds bracketing the two times
    msec = time2.Subtract(time1).TotalMilliseconds

  Loop Until msec < 500

  'Return the time of the event
  Return time3

End Function

End Class
```

Each event calls the same DailyEvent() function, passing a value to indicate which event is desired. Values such as 1, 2, or 3 could have been used to indicate the desired event, but by using enum to define a set of constants, these self-documenting indicators are much easier to understand.

The DailyEvent() function uses the same binary search for all three cases. Details are modified for the indicated search. For example, for the sunrise event the starting times for the binary search wrap the search around the morning hours, and the sunset search is centered around the evening hours. In both cases, the search narrows in on the desired sun elevation of -.8333 degrees, with the calculated refraction removed from the result.

The binary search avoids simply searching for values less than or greater than the target value. For example, the time of transit can use a search for the moment in time when the sun's always-increasing azimuth crosses from less than 180 degrees to greater than 180 degrees as it moves from east to west in the southern sky. However, this approach fails when the sun crosses the noon sky to the observer's north, as is the case in Australia, and even for points in the northern hemisphere between the equator and the Tropic of Cancer during part of the year. By comparing the sign of the sun's azimuth minus 180 degrees, these problems are avoided and the time of transit is found in all cases.

For readers tasked with translation of these routines for finding these daily events, take a close look at how the Date variables time1 and time2 are used. These two moments in time are first set with values for the type of event being searched for, and for the date in question. The logic is simple, just find a time half way between these two times with each step of the binary search. If the target programming language provides reasonable time calculation functionality, as does the .Net Framework, this task is easy. However, some creative programming may be required if the user has only integer variables for year, month, day, hour, minute, and second values for each moment in time. As a suggested approach, consider that the number of seconds into a day can be calculated as second + minute * 60 + hour * 3600.

SunPositionHighRes Class - High Accuracy Sun Position

The high accuracy sun position class that follows is very similar to the low accuracy one, except for all the extra detail that's been added in the main calculations. Either type-in or copy-and-paste the code blocks in the order presented to build the class module named SunPositionHighRes. Add this class to the main project as described earlier in this book, and verify the results of each step of the calculations using the numeric values listed below.

The accuracy of this algorithm is better than 0.00003 degrees. To get to this accuracy, more input data is required, including an accurate estimate of DeltaT in seconds, height of the observer's location above sea level in meters, barometric air pressure in millibars, and air temperature in degrees C. The weakest link in the accuracy is in the atmospheric refraction correction, as the best algorithms for estimating this adjustment are still just good estimates. Factors such as air temperature with elevation, amount of humidity in the air at various elevations, and so on all cause the refraction to vary a significant amount. The refraction algorithm presented here is used by experts as one of the best available.

You may copy-and-paste the following code block from
http://vb-book.com/source/EQBGDP.php

```
Imports System.Math

Public Class SunPositionHighRes

'Inputs
Public Property Year As Integer
Public Property Month As Integer
Public Property Day As Integer
Public Property Hour As Integer
Public Property Minute As Integer
Public Property Second As Integer
Public Property Timezone As Double
Public Property DeltaT As Double
Public Property Millibars As Double
Public Property DegC As Double
Public Property SiteElevation As Double
Public Property Latitude As Double
Public Property Longitude As Double

'Outputs
Public Property Azimuth As Double
Public Property Elevation As Double
Public Property Refraction As Double

'Constants
Const RadPerDeg As Double = PI / 180
Const DegPerRad As Double = 180 / PI

Private Enum rst As Integer
   sunrise
   sunset
   transit
End Enum
```

The first part of the SunPositionHighRes class imports the Math class, so all the trigonometric functions can be expressed with shorter notation (i.e. Sin(x) instead of Math.Sin(x)). The input properties are identical to those in the lower accuracy class, with the exception of a few new variables as mentioned above. Output properties and the enum constants are the same as before. The constants for conversions between radians and degrees are declared outside of the class methods because, unlike in the lower accuracy class, these constants are used in several methods instead of just in one. By declaring them at the class level, they are available throughout the class.

You may copy-and-paste the following code block from
http://vb-book.com/source/TDUQTR.php

```
Public Sub SunPos()

   'Local variables
   Dim Jd, Jde, Jc, Jce, Jme As Double
   Dim L, B, R As Double
   Dim theta, beta As Double
   Dim x0, x1, x2, x3, x4 As Double
   Dim nutLong, nutObli As Double
   Dim meanObli, u, trueObli As Double
   Dim aberration, lambda As Double
   Dim t1, t2, t3, t4, t5, t6, t7, t8 As Double
   Dim t9, t10, t11, t12, t13, t14, t15 As Double
   Dim t16, t17, t18, t19, t20, t21 As Double
   Dim rasc, decl As Double
   Dim meanSidereal, sidereal, H As Double
   Dim polrad, equrad, radrat As Double
   Dim ehpar, tu, tx, ty As Double
   Dim xx, yy As Double
   Dim parasc, torasc, todecl, toH As Double
   Dim toE, toA As Double

   'If DeltaT is not provided then approximate it
   If _DeltaT = 0 Then
      _DeltaT = ApproxDeltaT(_Year, _Month)
   End If

   'Julian day
   Jd = JulianDay(_Year, _Month, _Day, _Hour, _Minute, _
_Second, _Timezone)

   'Julian day ephemeris
   Jde = Jd + _DeltaT / 86400.0

   'Julian Century
   Jc = (Jd - 2451545) / 36525

   'Julian century ephemeris
   Jce = (Jde - 2451545) / 36525

   'Julian millennium ephemeris
   Jme = Jce / 10

   'Earth's position along the ecliptic
   L = EclipticLongitude(Jme)
   B = EclipticLatitude(Jme)
   R = SolarDistance(Jme)

   'Convert to geocentric longitude and latitude for sun
   theta = Range(L + 180, 0, 360)
   beta = -B

   'Mean elongation of the moon from the sun
   x0 = 297.85036 + 445267.11148 * Jce - 0.0019142 * Jce ^ 2
```

```
+ Jce ^ 3 / 189474

'Mean anomaly of the sun (earth)
x1 = 357.52772 + 35999.05034 * Jce - 0.0001603 * Jce ^ 2 -
Jce ^ 3 / 300000

'Mean anomaly of the moon
x2 = 134.96298 + 477198.867398 * Jce + 0.0086972 * Jce ^ 2
+ Jce ^ 3 / 56250

'Moon's argument of latitude
x3 = 93.27191 + 483202.017538 * Jce - 0.0036825 * Jce ^ 2
+ Jce ^ 3 / 327270

'Longitude of ascending node of moon's mean orbit
x4 = 125.04452 - 1934.136261 * Jce + 0.0020708 * Jce ^ 2 +
Jce ^ 3 / 450000

'Nutation
nutLong = NutationLongitude(Jce, x0, x1, x2, x3, x4)
nutObli = NutationObliquity(Jce, x0, x1, x2, x3, x4)

'Mean obliquity of the ecliptic
u = Jme / 10
meanObli = 84381.448 - 4680.93 * u - 1.55 * u ^ 2 +
1999.25 * u ^ 3 -
   51.38 * u ^ 4 - 249.67 * u ^ 5 - 39.05 * u ^ 6 + 7.12 *
u ^ 7 +
   27.87 * u ^ 8 + 5.79 * u ^ 9 + 2.45 * u ^ 10

'True obliquity of the ecliptic
trueObli = meanObli / 3600 + nutObli

'Aberration correction
aberration = -20.4898 / 3600 / R

'Apparent sun longitude
lambda = theta + nutLong + aberration

'Trig functions on various angular values
t1 = Sin(lambda * RadPerDeg)
t2 = Cos(lambda * RadPerDeg)
t3 = Sin(trueObli * RadPerDeg)
t4 = Cos(trueObli * RadPerDeg)
t5 = Sin(beta * RadPerDeg)
t6 = Cos(beta * RadPerDeg)
t7 = Tan(beta * RadPerDeg)
t8 = Sin(_Latitude * RadPerDeg)
t9 = Cos(_Latitude * RadPerDeg)
t10 = Tan(_Latitude * RadPerDeg)

'Geocentric right ascension and declination in radians
rasc = Atan2(t1 * t4 - t7 * t3, t2)
decl = Asin(t5 * t4 + t6 * t3 * t1)

'Convert right ascension and declination to degrees
rasc = Range(rasc * DegPerRad, 0, 360)
decl = Range(decl * DegPerRad, -180, 180)
```

```
'Mean sidereal time at Greenwich
meanSidereal = SiderealGreenwich(Jd)

'Apparent sidereal time at Greenwich
sidereal = Range(meanSidereal + nutLong * t4, 0, 360)

'Local hour angle
H = sidereal - _Longitude - rasc

'Equatorial horizontal parallax of the sun
ehpar = 8.794 / 3600 / R

'Earth's ratio of polar and equatorial radiuses
polrad = 6356755
equrad = 6378140
radrat = polrad / equrad

'Terms U, X and Y
tu = Atan(radrat * t10)
tx = Cos(tu) + (_SiteElevation * t9) / equrad
ty = radrat * Sin(tu) + (_SiteElevation * t8) / equrad

'Minimize occurrences of a few more trig functions
t11 = Sin(ehpar * RadPerDeg)
t12 = Sin(decl * RadPerDeg)
t13 = Cos(decl * RadPerDeg)
t14 = Sin(H * RadPerDeg)
t15 = Cos(H * RadPerDeg)

'Parallax in the sun's right ascension
yy = -tx * t11 * t14
xx = t13 - tx * t11 * t15
parasc = Range(Atan2(yy, xx) * DegPerRad, 0, 360)

'Trig function
t16 = Cos(parasc * RadPerDeg)

'Topocentric sun right ascension
torasc = Range(rasc + parasc, 0, 360)

'Topocentric sun declination
yy = (t12 - ty * t11) * t16
xx = t13 - tx * t11 * t15
todecl = Range(Atan2(yy, xx) * DegPerRad, -180, 180)

'Topocentric hour angle
toH = Range(H - parasc, 0, 360)

'A few more trig functions
t17 = Sin(todecl * RadPerDeg)
t18 = Cos(todecl * RadPerDeg)
t19 = Tan(todecl * RadPerDeg)
t20 = Sin(toH * RadPerDeg)
t21 = Cos(toH * RadPerDeg)

'Topocentric elevation angle
toE = Asin(t8 * t17 + t9 * t18 * t21)
toE = Range(toE * DegPerRad, -180, 180)
```

```
'Adjust elevation for refraction
_Refraction = RefractionCorrection(toE, _Millibars, _DegC)
_Elevation = toE + _Refraction

'Topocentric azimuth angle
yy = t20
xx = t21 * t8 - t19 * t9
toA = Range(Atan2(yy, xx) * DegPerRad, -180, 180)

'Convert from astronomer's azimuth to solar engineering
azimuth
_Azimuth = Range(toA + 180, 0, 360)

End Sub
```

The main SunPos() method has quite a few extra intermediate variables declared, and there are more steps in the calculations. Here are some words of explanation along with numerical values for verification of the code at each step.

DeltaT, the difference in time between Universal Time and Terrestrial Time, is either input by the user using data from the Internet, or it is estimated using the best estimating curves published at the time this book was written. The function ApproxDeltaT() is called to estimate DeltaT for any given year and month throughout history or in the future.

```
_DeltaT 67.0
```

Julian day, Julian day ephemeris, Julian century, Julian century ephemeris, and Julian millennium ephemeris are variables for the exact date and time in question. The Julian date is used in astronomical calculations because it provides a continuous sequence of day numbers that ignore the lumpiness of our civilian calendars. If two Julian day numbers vary by 1000, then there are exactly 1000 days between those two dates. The ephemeris versions of these values are adjusted for celestial events, which follow a time line that is smooth and ignores leap second adjustments. Some of the standard formulas use the exact decimal fraction of the centuries since the year 2000, and some formulas use the fraction of the millenium since the year 2000, hence the variable Jme being exactly 1/10 that of Jce.

```
Jd 2456283.3001388889
Jde 2456283.3009143518
Jc 0.12972758764925066
Jce 0.12972760888026907
Jme 0.012972760888026907
```

At the core of this algorithm are public tables of numbers for the VSOP87 theory, published by P. Bretagnon and G. Francou at the Bureau des Longitudes. These numbers provide a way to calculate, with a high degree of accuracy, the ecliptic longitude, ecliptic latitude, and distance of the earth from the sun at any moment in time. They published tables for each of the planets, including the earth, and for this algorithm the sun's longitude relative to the earth is found by simply adding 180 degrees to the earth's longitude relative to the sun. Similarly, the ecliptic latitude of the sun is found by simply changing the sign of the earth's ecliptic latitude. The distance of the sun from the earth is expressed in terms of the average orbital distance of the earth from the sun, a number that is always fairly close to 1. Part of each year R will be less than 1 and part of the year it will be greater than 1, because the earth's orbit is an ellipse intead of a perfect circle.

L, B, and R are the VSOP87 numbers calculated for the earth.

```
L 90.341765142750774
B -0.00014860488681733841
R 0.98369121271516935
```

Theta and beta are the geocentric longitude of the sun after the described adjustments, in degrees.

```
theta 270.34176514275077
beta 0.00014860488681733841
```

The earth spins in space with a wobble, much like a slightly unsteady toy top wiggles as it spins. In order to calculate the wobble angles for any given moment in time, several astronomical parameters for the earth and the moon must first be calculated. These include the mean anomaly of the earth and moon, the longitude of the ascending node of the moon's mean orbit, and so on. To get a firm grasp on what these terms mean, search for them on the Internet, or read about them in the glossary of a good astronomy text book. Here we just need to know they are required values for calculating the earth's nutation in latitude and nutation in obliquity, or in other words its wobble angles.

```
x0 58061.288013121579
x1 5027.59843986363
x2 62040.831154321306
x3 62777.91418935953
x4 -125.86631753328635
```

The nutation values are calculated in separate functions listed later on. The various parameters just described are passed to these functions to get an accurate value for the nutation angles.

```
nutLong 0.0038985829916783656
nutObli -0.0016379590834553566
```

The mean obliquity of the ecliptic is the tilt of the earth's axis relative to its orbit around the sun. This is the tilt that gives us the seasons, half the year tilting the northern hemisphere towards the sun, and half the year tilting away from the sun. The angle of tilt is approximately 23 1/2 degrees, but for these calculations a much more precise value is required. The formula presented here calculates the tilt as a function of time, as the tilt is always changing ever so slightly.

```
u 0.0012972760888026907
meanObli 84375.375543193775
```

The mean tilt of the earth is then adjusted for the nutation calculated previously to find the true obliquity of the ecliptic.

```
trueObli 23.43596635847037
```

Aberration is the slight shifted angle that sunlight strikes the earth due to the motion of the earth. Just like rain strikes your face at an angle as you walk forward in the rain, so does sunlight strike the earth at a slight angle as the earth moves forward in its orbit. The speed of light is much greater than the speed of the earth, so this angle is very small. Nevertheless, for the highest accuracy estimate of apparent sun position it must be taken into account.

```
aberration -0.0057859733192098094
```

The value of aberration, along with the nutation in longitude, are then used to adjust the apparent longitude of the sun relative to the earth.

```
lambda 270.33987775242321
```

Variables t1 through t10 are intermediate values of various trigonometric functions of the angles found so far. Each of these angles is first converted to radians before the trig function is applied. Most of the resulting values are used more than once, so isolating these calculations makes the followup calculations more efficient.

```
t1 -0.999982405823625
t2 0.0059319510445542947
t3 0.39772391572287641
t4 0.91750514268970851
t5 0.0000025936445595131647
t6 0.99999999999663647
t7 0.0000025936445595218883
t8 0.65081102730725138
t9 0.75923975576577918
t10 0.8571877623173656
```

The right ascension and declination of the sun is then calculated using these intermediate values. Right ascension and declination is how astronomers map out the exact position of stars and planets against the "fixed" background of the star field. Most stars basically have a fixed coordinate location in space (some neighboring stars are slowly drifting along in one direction or another) so their right ascension and declination are constant. The sun is in constant motion however, following a complete circle in right ascension on an annual basis.

The first set of rasc and decl values are in radians, and the second set is in degrees. The Range() function, described in more detail earlier in this books, is used to convert the radians into degrees using the appropriate range.

```
rasc -1.5643310048970345
decl -0.4090245559223713
rasc 270.370435657941
decl -23.43538077156461
```

Sidereal time is what astronomers use because it is based on the spin of the earth relative to the stars rather than on the earth's spin relative to the sun. A day, from noon to the next noon, is measured between the exact moments that the sun transits an oberver's meridian, or crosses the exact noon point in the sky. A sidereal day, on the other hand, is measured between the exact moments that a star crosses the observer's meridian. Over the course of a year, the difference between the lengths of the two days accumulates to a full day. A sidereal day is 23 hours, 56 minutes, 4.091 seconds in length. To ease further calculations, sidereal time is often expressed in degrees, as it is here.

```
meanSidereal 18.803677482763305
```

Apparent sidereal time at Greenwich is the mean sidereal time adjusted slightly for the earth's nutation, or wobble.

```
sidereal 18.807254452707355
```

Local hour angle is sidereal time, or angle at Greenwich, with the longitude of the observer worked in to shift the apparent sky rotation, and with the right ascension of the observed object worked in as well.

```
H -356.30484790523366
```

A few more subtle adjustments are required for the extreme accuracy of this algorithm. The angle to the sun's apparent position is shifted by parallax slightly, as a function of its distance from the earth. If the sun is straight overhead, at the zenith, there's no parallactic shift in its position. When the sun is near the horizon, the radius of the earth shifts the angle of the sun slightly. The earth's radius is much less than the distance to the sun, so the parallactic shift angle is very small.

```
ehpar 0.0024832770143745214
```

The earth is not a sphere. It's an oblate spheroid, due to the centrifugal force of its spin. As a result, "straight down" towards the center of the earth varies as a function of an observer's latitude, and hence the elevation angle to astronomical objects varies slightly too. The next set of calculations makes the necessary corrections for these small observation angles.

```
polrad 6356755.0
equrad 6378140.0
radrat 0.99664714164317492
tu 0.706993083747234
tx 0.76049927263673989
ty 0.64752764154094733
t11 0.000043341360126360689
t12 -.39771453841459542
t13 0.9175092075470771
t14 0.064447872477422868
t15 0.99792107490178494
yy -0.000002124271019832171l
xx 0.91747631499782756
parasc 359.99986734070188
```

The topocentric right ascension and declination of the sun are computed next. These angles are adjusted slightly from the results of the previous calculations.

```
t16 0.99999999999731959
torasc 270.37030299864296
yy -0.39774260314223314
xx 0.91747631499782756
todecl -23.437605694425656
```

The topocentric hour angle is computed, in preparation for the final calculation of elevation and azimuth of the sun's apparent position in the sky.

```
toH 3.6952847540644598
```

As before, trigonometric functions are calculated once for efficiency, with angles first converted to radians.

```
t17 -0.39775016704286287
t18 0.91749376271306315
t19 -0.43351811555285219
t20 0.064450183005364162
t21 0.997920925680274
```

Topocentric elevation is computed, first in radians, then converted to degrees in the range -180 to +180 degrees. The elevation angle won't ever be greater in magnitude than 90 degrees, but the range conversion should use a whole circle.

```
toE 0.45147060691316743
toE 25.867360350334366
```

The correction to elevation for atmospheric refraction is calculated and stored in its public property. If the elevation angle with no correction for refraction is reuqired, the user should subtract _Refraction from _Elevation. This technique is used internally by the class while doing a binary search for the times of sunrise and sunset.

```
_Refraction 0.02961943360104758
_Elevation 25.896979783935414
```

A related set of coordinate conversions is then used to calculate the azimuth angle of the sun. This topocentric azimuth angle is converted from radians to degrees in the range 0 to 360.

```
yy 0.064450183005364162
xx 0.97860213098577076
toA 3.7680260346098748
_Azimuth 183.76802603460987
```

Supporting Procedures

The high accuracy class has several more supporting procedures than did the low accuracy class. Type in or copy-and-paste these code blocks, in the order presented, into the SunPositionHighRes class module. A few words of explanation follow each procedure, and it includes example numerical values using the same example case as above.

You may copy-and-paste the following code block from
http://vb-book.com/source/LHHTLA.php

```
Private Function SiderealGreenwich(ByVal Jd As Double) As
Double

    'Local variables
    Dim stime, t1, t2, t3 As Double

    'Constants used in calculations
    Const c1 As Double = 280.46061837
    Const c2 As Double = 360.98564736629
    Const c3 As Double = 2451545.0
    Const c4 As Double = 0.000387933
    Const c5 As Double = 38710000.0
    Const c6 As Double = 36525.0

    'Calculate mean sidereal time at Greenwich
    t1 = (Jd - c3) / c6
    t2 = t1 * t1
    t3 = t2 * t1
    stime = (c1 + c2 * (Jd - c3) + c4 * t2 - t3 / c5) Mod 360
    Return stime
End Function
```

The SiderealGreenwich function calculates the Julian day number for a given date and time. The fractional part of Jd indicates the exact fractional part of the Julian date. The sidereal time is returned as a sidereal time angle as a Double precision variable.

Here are sample numerical values for the variables in this function. Use breakpoints, or simple output statements such as MsgBox() to verify proper operation of this code.

```
Jd 2456283.3001388889
t1 0.12972758764925066
t2 0.01682924699729401
t3 0.0021832176149123471
stime 18.803677482763305
```

You may copy-and-paste the following code block from
http://vb-book.com/source/PZGQZX.php

```
Private Function RefractionCorrection(ByVal elev As Double,
ByVal mbar As Double, ByVal degc As Double) As Double

    'Local variables
    Dim trmp, trmc, targ, trme, refr As Double

    'Term involving air pressure in millibars
    trmp = mbar / 1010

    'Term involving air temperature in deg C
    trmc = 283 / (273 + degc)

    'Convert the tan argument to radians
    targ = (elev + (10.3 / (elev + 5.11))) * RadPerDeg

    'Term for elevation shift
    trme = 1.02 / (60 * Tan(targ))

    'Total refraction shift
    refr = trme * trmp * trmc

    Return refr
End Function
```

The RefractionCorrection() function presented here takes into account the local site elevation, or height above sea level, the barometric pressure in millibars, and the air temperature in degrees C. These factors help adjust for the varying index of refraction of the air. Cold, dense air has a higher index of refraction, and there's more atmosphere to contend with for sites closer to sea level. Refraction of the atmospher is tricky to calculate accurately, but this algorithm does as good a job as any of the available models.

Here are sample numerical values for the variables in this function. Use breakpoints, or simple output statements such as MsgBox() to verify proper operation of this code.

```
elev 25.867360350334366
mbar 820.0
degc -5.0
trmp 0.81188118811881194
trmc 1.0559701492537314
targ 0.45727384227352019
trme 0.034548773106660165
refr 0.02961943360104758
```

You may copy-and-paste the following code block from

http://vb-book.com/source/VGQWII.php

```
Private Function ApproxDeltaT(ByVal theYear As Integer, ByVal
theMonth As Integer) As Double
  Dim y, t, u, dt As Double
  y = theYear + (theMonth - 0.5) / 12
  Select Case theYear
    Case -500 To 499
      u = y / 100
      dt = 10583.6 - 1014.41 * u + 33.78311 * u ^ 2 -
5.952053 * u ^ 3 -
        -0.1798452 * u ^ 4 + 0.022174192 * u ^ 5 +
0.0090316521 * u ^ 6
    Case 500 To 1599
      u = (y - 1000) / 100
      dt = 1574.2 - 556.01 * u + 71.23472 * u ^ 2 +
0.319781 * u ^ 3 -
        0.8503463 * u ^ 4 - 0.005050998 * u ^ 5 +
0.0083572073 * u ^ 6
    Case 1600 To 1699
      t = y - 1600
      dt = 120 - 0.9808 * t - 0.01532 * t ^ 2 + t ^ 3 /
7129
    Case 1700 To 1799
      t = y - 1700
      dt = 8.83 + 0.1603 * t - 0.0059285 * t ^ 2 +
0.00013336 * t ^ 3 - t ^ 4 / 1174000
    Case 1800 To 1859
      t = y - 1800
      dt = 13.72 - 0.332447 * t + 0.0068612 * t ^ 2 +
```

```
0.0041116 * t ^ 3 - 0.00037436 * t ^ 4 +
      0.0000121272 * t ^ 5 - 0.0000001699 * t ^ 6 +
0.000000000875 * t ^ 7
    Case 1860 To 1899
      t = y - 1860
      dt = 7.62 + 0.5737 * t - 0.251754 * t ^ 2 +
0.01680668 * t ^ 3 -
        0.0004473624 * t ^ 4 + t ^ 5 / 233174
    Case 1900 To 1919
      t = y - 1900
      dt = -2.79 + 1.494119 * t - 0.0598939 * t ^ 2 +
0.0061966 * t ^ 3 - 0.000197 * t ^ 4
    Case 1920 To 1940
      t = y - 1920
      dt = 21.2 + 0.84493 * t - 0.0761 * t ^ 2 + 0.0020936
* t ^ 3
    Case 1941 To 1960
      t = y - 1950
      dt = 29.07 + 0.407 * t - t ^ 2 / 233 + t ^ 3 / 2547
    Case 1961 To 1985
      t = y - 1975
      dt = 45.45 + 1.067 * t - t ^ 2 / 260 - t ^ 3 / 718
    Case 1986 To 2004
      t = y - 2000
      dt = 63.86 + 0.3345 * t - 0.060374 * t ^ 2 +
0.0017275 * t ^ 3 +
        0.000651814 * t ^ 4 + 0.00002373599 * t ^ 5
    Case 2005 To 2049
      t = y - 2000
      dt = 62.92 + 0.32217 * t + 0.005589 * t ^ 2
    Case 2050 To 2149
      dt = -20 + 32 * ((y - 1820) / 100) ^ 2 - 0.5628 *
(2150 - y)
    Case Else
      t = (theYear - 1820) / 100
      dt = -20 + 32 * t ^ 2
  End Select
  Return dt
End Function
```

The most accurate way to obtain DeltaT values for the current month is on the Internet. A very good resource for understanding DeltaT is at
http://eclipse.gsfc.nasa.gov/SEhelp/deltaT.html

Monthly updates for DeltaT are available at
http://maia.usno.navy.mil/ser7/deltat.data

The rather lengthy function provided here returns historical values for DeltaT based on a variety of polynomial approximations. It also attempts to predict future values of DeltaT, although this is a tricky business since the earth's spin is somewhat unpredictable. DeltaT is the difference in time of celestial events versus earth-sun events, and as such its value doesn't have to be perfect. The various calculations in this class that require DeltaT will return very small variations if DeltaT is off by a matter of seconds. It's not a big deal, although for the extreme accuracy this class is shooting for, using an accurate value for the current DeltaT is a good policy.

You may copy-and-paste the following code block from
http://vb-book.com/source/OISHPX.php

```
Private Function JulianDay(ByVal ye As Integer, ByVal mo As
Integer, ByVal da As Integer,
ByVal ho As Integer, ByVal mi As Integer, ByVal se As
Integer, ByVal tz As Double) As Double

   'Local variables
   Dim daynum, tmp, jd As Double

   'Decimal fraction of the day
   daynum = da + (ho + tz + (mi + se / 60) / 60) / 24

   'Adjust month and year if needed
   If mo < 3 Then
      mo += 12
      ye -= 1
   End If

   'Julian day number
   jd = Floor(365.25 * (ye + 4716)) + Floor(30.6001 * (mo +
1)) + daynum - 1524.5

   'Adjust from Julian to Gregorian calendar after Oct 4,
1582
   If jd > 2299160 Then
      tmp = Floor(ye / 100)
      jd += (2 - Floor(ye / 100) + Floor(tmp / 4))
   End If

   Return jd
End Function
```

The JulianDay function converts a local date and time, plus the local time zone offest from Greenwich, to calculate the Julian day number. This number is used by astronomers for a wide variety of calculations of astronomical events. Sometimes, a modified Julian day number is used, such as in the low accuracy sun position class presented earlier in this book, but here the full value is calculated. As shown below, the number of days since Julian day number 0 is considerable. In this case, December 21 of 2012 converts to a Julian day number just under 2 1/2 million in size. Julian day number zero is way back in time, at noon of January 1, 4713 BC.

```
ye 2012
mo 12
da 21
ho 12
mi 12
se 12
tz 7.0
daynum 21.800138888888888
jd 2456296.3001388889
tmp 20.0
jd 2456283.3001388889
```

You may copy-and-paste the following code block from

http://vb-book.com/source/EFTIAQ.php

```
Private Function EclipticLongitude(ByVal Jme As Double) As
Double
  Dim L0, L1, L2, L3, L4, L5, Lt, Ld As Double
  L0 = SumTerms(aryL0, Jme)
  L1 = SumTerms(aryL1, Jme)
  L2 = SumTerms(aryL2, Jme)
  L3 = SumTerms(aryL3, Jme)
  L4 = SumTerms(aryL4, Jme)
  L5 = SumTerms(aryL5, Jme)
  Lt = L0
  Lt += L1 * Jme
  Lt += L2 * Jme ^ 2
  Lt += L3 * Jme ^ 3
  Lt += L4 * Jme ^ 4
  Lt += L5 * Jme ^ 5
  Ld = Range((Lt / 100000000.0) * DegPerRad, 0, 360)
  Return Ld
End Function
```

Ecliptic longitude is a measure, in total angular degrees, of the total number of degrees the earth has orbited around the sun at any given moment in time. This function repeatedly calls SumTerms() to form sums using the appropriate numbers from the VSOP87 tables.

```
Jme 0.012972760888026907
L0 174618838.41581953
L1 628331798202.32153
L2 58676.331517331979
L3 279.66831291745387
L4 -120.41999142411845
L5 -0.9999987317275395
Lt 8325817024.7139063
Ld 90.341765142750774
```

You may copy-and-paste the following code block from

http://vb-book.com/source/VTQYIP.php

```
Private Function EclipticLatitude(ByVal Jme As Double) As
Double
    Dim B0, B1, Bt, Bd As Double
    B0 = SumTerms(aryB0, Jme)
    B1 = SumTerms(aryB1, Jme)
    Bt = B0 + B1 * Jme
    Bd = Range((Bt / 100000000.0) * DegPerRad, -180, 180)
    Return Bd
End Function
```

Ecliptic latitude of the earth is also calculated by summing many terms using the VSOP87 tables. The final result is an angle in radians that is then converted to degrees latitude ranging from -90 to +90 degrees. As the example numerical results show, the ecliptic latitude of the earth is a very small angle. This makes sense as the earth generally orbits the sun in the plane of the ecliptic.

```
Jme 0.012972760888026907
B0 -259.55332992268575
B1 14.559265590277853
Bt -259.36445605147782
Bd -0.00014860488681733841
```

You may copy-and-paste the following code block from
http://vb-book.com/source/JEASRT.php

```
Private Function SolarDistance(ByVal Jme As Double) As Double
   Dim R0, R1, R2, R3, R4, Rt As Double
   R0 = SumTerms(aryR0, Jme)
   R1 = SumTerms(aryR1, Jme)
   R2 = SumTerms(aryR2, Jme)
   R3 = SumTerms(aryR3, Jme)
   R4 = SumTerms(aryR4, Jme)
   Rt = R0
   Rt += R1 * Jme
   Rt += R2 * Jme ^ 2
   Rt += R3 * Jme ^ 3
   Rt += R4 * Jme ^ 4
   Rt /= 100000000.0
   Return Rt
End Function
```

Solar distance is the distance between the earth and the sun, expressed in standard astronomical units where one AU is 92,955,887.6 miles. The earth's orbit is slightly elliptical, so the distance to the sun varies throughout the year, but the value of SolarDistance is always close to 1, in this case about 0.9837 AU.

```
Jme 0.012972760888026907
R0 98368318.812072426
R1 61812.196558674666
R2 3474.8685798955876
R3 -89.651265159698752
R4 -2.9155324058221668
Rt 0.98369121271516935
```

You may copy-and-paste the following code block from
http://vb-book.com/source/KAJOCU.php

```
Private Function SumTerms(ByVal Ary(,) As Double, ByVal Jme
As Double) As Double
   Dim i As Integer
   Dim Sum As Double
   For i = 0 To UBound(Ary, 1)
      Sum += Ary(i, 0) * Cos(Ary(i, 1) + Ary(i, 2) * Jme)
   Next
   Return Sum
End Function
```

The SumTerms function is the workhorse routine that sums many terms from the VSOP87 tables to calculate the earth's ecliptic longitude, latitude, and disance for any moment in time. As shown below, the tables are presented as many rows of three numbers. These three numbers are processed one row at a time and then added together to calculate an angle in radians, or a distance in Astronomical Units. No sample numbers are provided for this function, as it's called repeatedly and does a lot of looping internally to form the sums. If values for L, B, and R are validated to match those listed above, then this function is working as intended.

You may copy-and-paste the following code block from
http://vb-book.com/source/IRRSWA.php

```
Private Function NutationLongitude(ByVal Jce As Double, ByVal
x0 As Double,
ByVal x1 As Double, ByVal x2 As Double, ByVal x3 As Double,
ByVal x4 As Double) As Double
  Dim i As Integer
  Dim x0r, x1r, x2r, x3r, x4r As Double
  Dim sumLon, sumXY As Double
  x0r = x0 * RadPerDeg
  x1r = x1 * RadPerDeg
  x2r = x2 * RadPerDeg
  x3r = x3 * RadPerDeg
  x4r = x4 * RadPerDeg
  For i = 0 To UBound(nut, 1)
    sumXY = x0r * nut(i, 0)
    sumXY += x1r * nut(i, 1)
    sumXY += x2r * nut(i, 2)
    sumXY += x3r * nut(i, 3)
    sumXY += x4r * nut(i, 4)
    sumLon += (nut(i, 5) + nut(i, 6) * Jce) * Sin(sumXY)
  Next
  sumLon /= 36000000
  Return sumLon
End Function
```

The NutationLongitude() function uses intermediate results calculated above to calculate the small wobble angle of the earth's axis for any given moment in time. The values of x0 through x4 are functions of the motions of both the earth and the moon relative to the sun.

```
Jce 0.12972760888026907
x0 58061.288013121579
x1 5027.59843986363
x2 62040.831154321306
x3 62777.91418935953
x4 -125.86631753328635
x0r 1013.3606437776882
x1r 87.748146243750483
x2r 1082.816774316781
x3r 1095.6813001387907
x4r -2.1967816583165143
sumLon 0.0038985829916783656
```

You may copy-and-paste the following code block from
http://vb-book.com/source/WDYNXF.php

```
Private Function NutationObliquity(ByVal Jce As Double, ByVal
x0 As Double,
ByVal x1 As Double, ByVal x2 As Double, ByVal x3 As Double,
ByVal x4 As Double) As Double
    Dim i As Integer
    Dim x0r, x1r, x2r, x3r, x4r As Double
    Dim sumObl, sumXY As Double
    x0r = x0 * RadPerDeg
    x1r = x1 * RadPerDeg
    x2r = x2 * RadPerDeg
    x3r = x3 * RadPerDeg
    x4r = x4 * RadPerDeg
    For i = 0 To UBound(nut, 1)
        sumXY = x0r * nut(i, 0)
        sumXY += x1r * nut(i, 1)
        sumXY += x2r * nut(i, 2)
        sumXY += x3r * nut(i, 3)
        sumXY += x4r * nut(i, 4)
        sumObl += (nut(i, 7) + nut(i, 8) * Jce) * Cos(sumXY)
    Next
    sumObl /= 36000000
    Return sumObl
End Function
```

The nutation in obliquity, or the earth's spin wobble relative to the plane of the ecliptic, is calculated in much the same way as the nutation in longitude. This function also uses x0 through x4 parameters that are functions of the motion of the earth and moon relative to the sun.

```
Jce 0.12972760888026907
x0 58061.288013121579
x1 5027.59843986363
x2 62040.831154321306
x3 62777.91418935953
x4 -125.86631753328635
x0r 1013.3606437776882
x1r 87.748146243750483
x2r 1082.816774316781
x3r 1095.6813001387907
x4r -2.1967816583165143
sumObl -0.0016379590834553566
```

You may copy-and-paste the following code block from
http://vb-book.com/source/IZRTDN.php

```
Private Sub SetDateTime(ByVal dt As Date)
    _Year = dt.Year
    _Month = dt.Month
    _Day = dt.Day
    _Hour = dt.Hour
    _Minute = dt.Minute
    _Second = dt.Second
End Sub
```

This function is identical to the one of the same name in the low accuracy sun position class. Its purpose is to extract values from a single Date variable to populate the date and time properties of the class. This is a private function, unavailable to external calling prodedures. It is used internally by the class during the binary searches for the times of sunrise, sunset, and transit. For more information about this prodedure refer to the description in the low accuracy class presented earlier.

You may copy-and-paste the following code block from
http://vb-book.com/source/QMGMAR.php

```
Private Function Range(ByVal X As Double, ByVal Rmin As
Double, ByVal Rmax As Double) As Double
  Dim Delta As Double = Rmax - Rmin
  Return (((X - Rmin) Mod Delta) + Delta) Mod Delta + Rmin
End Function
```

The Range() function is also described in the low accuracy class presented earlier. Its purpose is to bring very large angles, whether in radians or degrees, into an appropriate range within one unit circle of angular measurement.

You may copy-and-paste the following code block from
http://vb-book.com/source/IIYJSB.php

```
Private aryL0(,) As Double = {
{175347046, 0, 0},
{3341656, 4.6692568, 6283.07585},
{34894, 4.6261, 12566.1517},
{3497, 2.7441, 5753.3849},
{3418, 2.8289, 3.5231},
{3136, 3.6277, 77713.7715},
{2676, 4.4181, 7860.4194},
{2343, 6.1352, 3930.2097},
{1324, 0.7425, 11506.7698},
{1273, 2.0371, 529.691},
{1199, 1.1096, 1577.3435},
{990, 5.233, 5884.927},
{902, 2.045, 26.298},
{857, 3.508, 398.149},
{780, 1.179, 5223.694},
{753, 2.533, 5507.553},
{505, 4.583, 18849.228},
{492, 4.205, 775.523},
{357, 2.92, 0.067},
{317, 5.849, 11790.629},
{284, 1.899, 796.298},
{271, 0.315, 10977.079},
{243, 0.345, 5486.778},
{206, 4.806, 2544.314},
{205, 1.869, 5573.143},
{202, 2.458, 6069.777},
{156, 0.833, 213.299},
{132, 3.411, 2942.463},
{126, 1.083, 20.775},
```

```
{115, 0.645, 0.98},
{103, 0.636, 4694.003},
{102, 0.976, 15720.839},
{102, 4.267, 7.114},
{99, 6.21, 2146.17},
{98, 0.68, 155.42},
{86, 5.98, 161000.69},
{85, 1.3, 6275.96},
{85, 3.67, 71430.7},
{80, 1.81, 17260.15},
{79, 3.04, 12036.46},
{75, 1.76, 5088.63},
{74, 3.5, 3154.69},
{74, 4.68, 801.82},
{70, 0.83, 9437.76},
{62, 3.98, 8827.39},
{61, 1.82, 7084.9},
{57, 2.78, 6286.6},
{56, 4.39, 14143.5},
{56, 3.47, 6279.55},
{52, 0.19, 12139.55},
{52, 1.33, 1748.02},
{51, 0.28, 5856.48},
{49, 0.49, 1194.45},
{41, 5.37, 8429.24},
{41, 2.4, 19651.05},
{39, 6.17, 10447.39},
{37, 6.04, 10213.29},
{37, 2.57, 1059.38},
{36, 1.71, 2352.87},
{36, 1.78, 6812.77},
{33, 0.59, 17789.85},
{30, 0.44, 83996.85},
{30, 2.74, 1349.87},
{25, 3.16, 4690.48}}

Private aryL1(,) As Double = {
{628331966747, 0, 0},
{206059, 2.678235, 6283.07585},
{4303, 2.6351, 12566.1517},
{425, 1.59, 3.523},
{119, 5.796, 26.298},
{109, 2.966, 1577.344},
{93, 2.59, 18849.23},
{72, 1.14, 529.69},
{68, 1.87, 398.15},
{67, 4.41, 5507.55},
{59, 2.89, 5223.69},
{56, 2.17, 155.42},
{45, 0.4, 796.3},
{36, 0.47, 775.52},
{29, 2.65, 7.11},
{21, 5.34, 0.98},
{19, 1.85, 5486.78},
{19, 4.97, 213.3},
{17, 2.99, 6275.96},
{16, 0.03, 2544.31},
{16, 1.43, 2146.17},
{15, 1.21, 10977.08},
```

```
{12,  2.83,  1748.02},
{12,  3.26,  5088.63},
{12,  5.27,  1194.45},
{12,  2.08,  4694},
{11,  0.77,  553.57},
{10,  1.3,  6286.6},
{10,  4.24,  1349.87},
{9,  2.7,  242.73},
{9,  5.64,  951.72},
{8,  5.3,  2352.87},
{6,  2.65,  9437.76},
{6,  4.67,  4690.48}}

Private aryL2(,) As Double = {
{52919, 0, 0},
{8720, 1.0721, 6283.0758},
{309, 0.867, 12566.152},
{27, 0.05, 3.52},
{16, 5.19, 26.3},
{16, 3.68, 155.42},
{10, 0.76, 18849.23},
{9, 2.06, 77713.77},
{7, 0.83, 775.52},
{5, 4.66, 1577.34},
{4, 1.03, 7.11},
{4, 3.44, 5573.14},
{3, 5.14, 796.3},
{3, 6.05, 5507.55},
{3, 1.19, 242.73},
{3, 6.12, 529.69},
{3, 0.31, 398.15},
{3, 2.28, 553.57},
{2, 4.38, 5223.69},
{2, 3.75, 0.98}}

Private aryL3(,) As Double = {
{289, 5.844, 6283.076},
{35, 0, 0},
{17, 5.49, 12566.15},
{3, 5.2, 155.42},
{1, 4.72, 3.52},
{1, 5.3, 18849.23},
{1, 5.97, 242.73}}

Private aryL4(,) As Double = {
{114, 3.142, 0},
{8, 4.13, 6283.08},
{1, 3.84, 12566.15}}

Private aryL5(,) As Double = {
{1, 3.14, 0}}

Private aryB0(,) As Double = {
{280, 3.199, 84334.662},
{102, 5.422, 5507.553},
{80, 3.88, 5223.69},
{44, 3.7, 2352.87},
{32, 4, 1577.34}}
```

```
Private aryB1(,) As Double = {
{9, 3.9, 5507.55},
{6, 1.73, 5223.69}}

Private aryR0(,) As Double = {
{100013989, 0, 0},
{1670700, 3.0984635, 6283.07585},
{13956, 3.05525, 12566.1517},
{3084, 5.1985, 77713.7715},
{1628, 1.1739, 5753.3849},
{1576, 2.8469, 7860.4194},
{925, 5.453, 11506.77},
{542, 4.564, 3930.21},
{472, 3.661, 5884.927},
{346, 0.964, 5507.553},
{329, 5.9, 5223.694},
{307, 0.299, 5573.143},
{243, 4.273, 11790.629},
{212, 5.847, 1577.344},
{186, 5.022, 10977.079},
{175, 3.012, 18849.228},
{110, 5.055, 5486.778},
{98, 0.89, 6069.78},
{86, 5.69, 15720.84},
{86, 1.27, 161000.69},
{65, 0.27, 17260.15},
{63, 0.92, 529.69},
{57, 2.01, 83996.85},
{56, 5.24, 71430.7},
{49, 3.25, 2544.31},
{47, 2.58, 775.52},
{45, 5.54, 9437.76},
{43, 6.01, 6275.96},
{39, 5.36, 4694},
{38, 2.39, 8827.39},
{37, 0.83, 19651.05},
{37, 4.9, 12139.55},
{36, 1.67, 12036.46},
{35, 1.84, 2942.46},
{33, 0.24, 7084.9},
{32, 0.18, 5088.63},
{32, 1.78, 398.15},
{28, 1.21, 6286.6},
{28, 1.9, 6279.55},
{26, 4.59, 10447.39}}

Private aryR1(,) As Double = {
{103019, 1.10749, 6283.07585},
{1721, 1.0644, 12566.1517},
{702, 3.142, 0},
{32, 1.02, 18849.23},
{31, 2.84, 5507.55},
{25, 1.32, 5223.69},
{18, 1.42, 1577.34},
{10, 5.91, 10977.08},
{9, 1.42, 6275.96},
{9, 0.27, 5486.78}}

Private aryR2(,) As Double = {
```

```
{4359, 5.7846, 6283.0758},
{124, 5.579, 12566.152},
{12, 3.14, 0},
{9, 3.63, 77713.77},
{6, 1.87, 5573.14},
{3, 5.47, 18849.23}}

Private aryR3(,) As Double = {
{145, 4.273, 6283.076},
{7, 3.92, 12566.15}}

Private aryR4(,) As Double = {
{4, 2.56, 6283.08}}

Private nut(,) As Double = {
{0, 0, 0, 0, 1, -171996, -174.2, 92025, 8.9},
{-2, 0, 0, 2, 2, -13187, -1.6, 5736, -3.1},
{0, 0, 0, 2, 2, -2274, -0.2, 977, -0.5},
{0, 0, 0, 0, 2, 2062, 0.2, -895, 0.5},
{0, 1, 0, 0, 0, 1426, -3.4, 54, -0.1},
{0, 0, 1, 0, 0, 712, 0.1, -7, 0},
{-2, 1, 0, 2, 2, -517, 1.2, 224, -0.6},
{0, 0, 0, 2, 1, -386, -0.4, 200, 0},
{0, 0, 1, 2, 2, -301, 0, 129, -0.1},
{-2, -1, 0, 2, 2, 217, -0.5, -95, 0.3},
{-2, 0, 1, 0, 0, -158, 0, 0, 0},
{-2, 0, 0, 2, 1, 129, 0.1, -70, 0},
{0, 0, -1, 2, 2, 123, 0, -53, 0},
{2, 0, 0, 0, 0, 63, 0, 0, 0},
{0, 0, 1, 0, 1, 63, 0.1, -33, 0},
{2, 0, -1, 2, 2, -59, 0, 26, 0},
{0, 0, -1, 0, 1, -58, -0.1, 32, 0},
{0, 0, 1, 2, 1, -51, 0, 27, 0},
{-2, 0, 2, 0, 0, 48, 0, 0, 0},
{0, 0, -2, 2, 1, 46, 0, -24, 0},
{2, 0, 0, 2, 2, -38, 0, 16, 0},
{0, 0, 2, 2, 2, -31, 0, 13, 0},
{0, 0, 2, 0, 0, 29, 0, 0, 0},
{-2, 0, 1, 2, 2, 29, 0, -12, 0},
{0, 0, 0, 2, 0, 26, 0, 0, 0},
{-2, 0, 0, 2, 0, -22, 0, 0, 0},
{0, 0, -1, 2, 1, 21, 0, -10, 0},
{0, 2, 0, 0, 0, 17, -0.1, 0, 0},
{2, 0, -1, 0, 1, 16, 0, -8, 0},
{-2, 2, 0, 2, 2, -16, 0.1, 7, 0},
{0, 1, 0, 0, 1, -15, 0, 9, 0},
{-2, 0, 1, 0, 1, -13, 0, 7, 0},
{0, -1, 0, 0, 1, -12, 0, 6, 0},
{0, 0, 2, -2, 0, 11, 0, 0, 0},
{2, 0, -1, 2, 1, -10, 0, 5, 0},
{2, 0, 1, 2, 2, -8, 0, 3, 0},
{0, 1, 0, 2, 2, 7, 0, -3, 0},
{-2, 1, 1, 0, 0, -7, 0, 0, 0},
{0, -1, 0, 2, 2, -7, 0, 3, 0},
{2, 0, 0, 2, 1, -7, 0, 3, 0},
{2, 0, 1, 0, 0, 6, 0, 0, 0},
{-2, 0, 2, 2, 2, 6, 0, -3, 0},
{-2, 0, 1, 2, 1, 6, 0, -3, 0},
{2, 0, -2, 0, 1, -6, 0, 3, 0},
```

```
{2, 0, 0, 0, 1, -6, 0, 3, 0},
{0, -1, 1, 0, 0, 5, 0, 0, 0},
{-2, -1, 0, 2, 1, -5, 0, 3, 0},
{-2, 0, 0, 0, 1, -5, 0, 3, 0},
{0, 0, 2, 2, 1, -5, 0, 3, 0},
{-2, 0, 2, 0, 1, 4, 0, 0, 0},
{-2, 1, 0, 2, 1, 4, 0, 0, 0},
{0, 0, 1, -2, 0, 4, 0, 0, 0},
{-1, 0, 1, 0, 0, -4, 0, 0, 0},
{-2, 1, 0, 0, 0, -4, 0, 0, 0},
{1, 0, 0, 0, 0, -4, 0, 0, 0},
{0, 0, 1, 2, 0, 3, 0, 0, 0},
{0, 0, -2, 2, 2, -3, 0, 0, 0},
{-1, -1, 1, 0, 0, -3, 0, 0, 0},
{0, 1, 1, 0, 0, -3, 0, 0, 0},
{0, -1, 1, 2, 2, -3, 0, 0, 0},
{2, -1, -1, 2, 2, -3, 0, 0, 0},
{0, 0, 3, 2, 2, -3, 0, 0, 0},
{2, -1, 0, 2, 2, -3, 0, 0, 0}}
```

These two dimensional arrays hold the tabular data values as published in the VSOP87 theory. Each array is processed by the SumTerms() function to calculate planetary longitude, latitude, and distance from the sun. The data used here is just for the earth, whereas the full set of tables repeats similar data items for each of the planets. For calculating sun position, only the position of the earth is required.

The data presented here is actually a shortened subset of the full tables. As explained by Jean Meeus in his valuable book titled Astronomical Algorithms, these shorter versions of the tables are plenty accurate enough for the high accuracy goals of the algorithm presented in this book.

You may copy-and-paste the following code block from
http://vb-book.com/source/XEXHVH.php

```
Public Function Sunrise() As Date
  Return (DailyEvent(rst.sunrise))
End Function

Public Function Sunset() As Date
  Return (DailyEvent(rst.sunset))
End Function
```

```
Public Function Transit() As Date
  Return (DailyEvent(rst.transit))
End Function
```

These three functions calculate the times of sunrise, sunset, and transit (solar noon) for any given date. Each calls the same function, named DailyEvents(), passing an enumerated contant that indicates which type of event is to be calculated using a binary search algorithm.

You may copy-and-paste the following code block from
http://vb-book.com/source/SDIPZO.php

```
Private Function DailyEvent(ByVal dayevent As rst) As Date

  'Local variables
  Dim time1, time2, time3 As Date
  Dim azim1, azim2 As Double
  Dim elev1, elev2 As Double
  Dim ms As Double
  Dim sec As Double

  'Select bracketing times for binary search
  Try
    Select Case dayevent
      Case rst.sunrise
        time1 = New Date(_Year, _Month, _Day, 0, 0, 1)
        time2 = New Date(_Year, _Month, _Day, 12, 0, 0)
      Case rst.sunset
        time1 = New Date(_Year, _Month, _Day, 12, 0, 0)
        time2 = New Date(_Year, _Month, _Day, 23, 59, 59)
      Case rst.transit
        time1 = New Date(_Year, _Month, _Day, 6, 0, 0)
        time2 = New Date(_Year, _Month, _Day, 18, 0, 0)
    End Select
  Catch
    _Azimuth = 0
    _Elevation = 0
    Return New Date(0)
  End Try

  'Calculate sun position for first time
  SetDateTime(time1)
  Me.SunPos()
  azim1 = _Azimuth
  elev1 = _Elevation

  'Calculate sun elevation for second time
  SetDateTime(time2)
  Me.SunPos()
```

```
azim2 = _Azimuth
elev2 = _Elevation

'Bail out if sun doesn't behave
Select Case dayevent
  Case rst.sunrise, rst.sunset
    If Sign(elev1) = Sign(elev2) Then
      Return New Date(0)
    End If
  Case rst.transit
    If Sign(180.0 - azim1) = Sign(180.0 - azim2) Then
      Return New Date(0)
    End If
End Select

Do
  'Average the two times
  ms = time2.Subtract(time1).TotalMilliseconds / 2
  time3 = time1.Add(TimeSpan.FromMilliseconds(ms))

  'Calulate sun position for new time
  SetDateTime(time3)
  Me.SunPos()

  'Determine which time, azimuth, and elevation to replace
  Select Case dayevent
    Case rst.sunrise
      If _Elevation - _Refraction + 0.8333 < 0.0 Then
        time1 = time3
        azim1 = _Azimuth
        elev1 = _Elevation
      Else
        time2 = time3
        azim2 = _Azimuth
        elev2 = _Elevation
      End If
    Case rst.sunset
      If _Elevation - _Refraction + 0.8333 > 0.0 Then
        time1 = time3
        azim1 = _Azimuth
        elev1 = _Elevation
      Else
        time2 = time3
        azim2 = _Azimuth
        elev2 = _Elevation
      End If
    Case rst.transit
      If Sign(180.0 - _Azimuth) = Sign(180.0 - azim1) Then
        time1 = time3
        azim1 = _Azimuth
        elev1 = _Elevation
      Else
        time2 = time3
        azim2 = _Azimuth
        elev2 = _Elevation
      End If
    End Select

  'Number of seconds between the two times
```

```
    sec = time2.Subtract(time1).TotalSeconds

  Loop Until sec <= 1

  'Return the time of the event
  Return time3

End Function

End Class
```

The DailyEvent() function uses a binary search technique to narrow in on either the exact time of sunrise, sunset, or transit (solar noon) for any given date. The search terminates when the binary search range is less than 1 second. This function is identical to the one of the same name presented in the low accuracy sun position class. Refer to its description earlier in this book for more information.

To verify proper operation, here are the values of the returned time3 variable for each type of event.

```
Sunrise()
time3.Ticks 634916711672490000

Sunset()
time3.Ticks 634917045229480000

Transit()
time3.Ticks 634916878447640000
```

Debugging

Every effort was made by the author to provide a solid working example of the calculations all through the code presented in this book. The value of every variable was presented for a single test case, so the user can verify that each line of code was accurately entered into their new program. This is something the author wishes he had when first programming sun position for several major solar energy sites a few years back. Hopefully this will ease the process greatly for those that follow in his footsteps to implement solar energy projects of all types.

As a word of caution, the most common cause of problems and hassles, other than simple typographical transcription errors, has been in the varied use of radians, degrees, minutes, and seconds in the literature when calculating the various astronomical parameters. This book uses both radians and degrees in the various calculations, so be aware of the proper units as they're used.

Every effort was made by the author to maintain the accuracy of all formulas and results. If any errors are noticed please report them to the author, and be sure to check for any corrections, updates, or notices at this web site address:

http://vb-book.com/source/SFFCHW.php

Appendices

Appendix A

Visual Basic Express Setup Guidelines

These instructions assume you'll be downloading and installing Visual Basic Express 2010. If a newer version of Visual Basic is available by the time you read this you might need to modify the instructions as appropriate.

Browse to:
`http://www.microsoft.com/express/Download s`

Click on "Visual Basic Express 2010" to download the installation program `vb_web.exe`

Double-click `vb_web.exe` and follow the installation instructions to install the complete package.

When installation is complete, Visual Basic Express is ready to rock and roll.

From the Start menu select *Microsoft Visual Studio 2010 Express -> Microsoft Visual Basic 2010 Express*

You'll need to "register" Visual Basic before 30 days are up, but there's no fee involved and the process is easy.

Appendix B

Example Calculations

The main program in this book was set up to demonstrate both the low and high accuracy sun position algorithms using three test cases. The Ault, Colorado test case is the primary example, with numerical results for all varaibles presented throughout the code descriptions in this book, Tow other test cases are shown in the main form as well, one for a day in 1981 at the Sandia Labs solar enrgy facility, and the other at the National Renewable Laboratory in Golden, Colorado.

Screen copies of each of these test case runs are shown here. To aid in readability, the left and right half of the main form is shown separately for each test case. The left half of the form shows the input values, and the right half shows the results of the calculations. The user should use the following figures to verify proper operation of their implementation of the code.

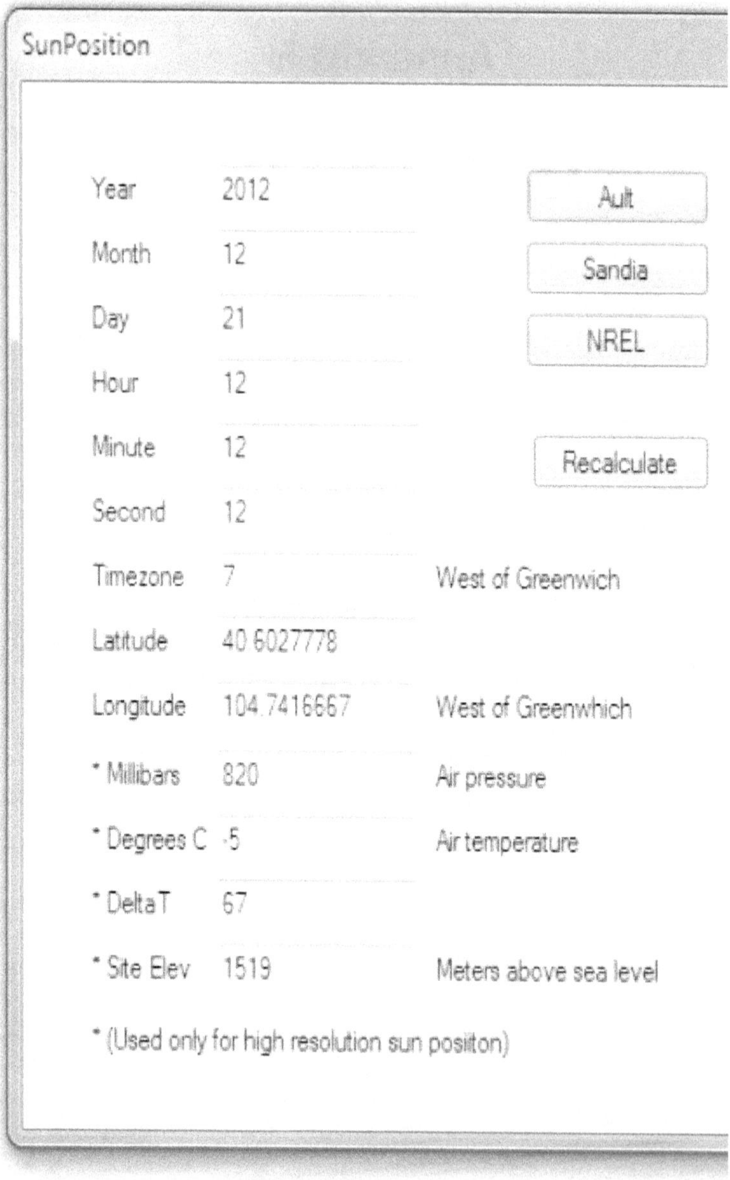

Year	2012		Ault
Month	12		Sandia
Day	21		NREL
Hour	12		
Minute	12		Recalculate
Second	12		
Timezone	7	West of Greenwich	
Latitude	40.6027778		
Longitude	104.7416667	West of Greenwhich	
* Millibars	820	Air pressure	
* Degrees C	-5	Air temperature	
* DeltaT	67		
* Site Elev	1519	Meters above sea level	

* (Used only for high resolution sun position)

Figure 4 - The input values for the Ault case

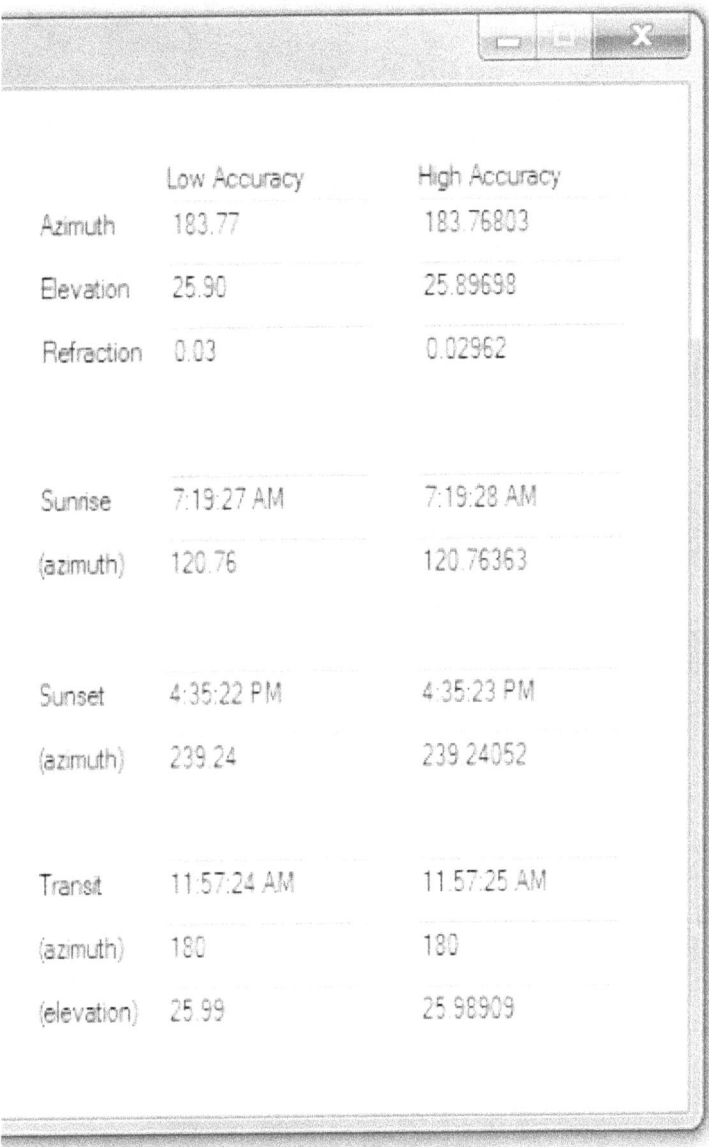

	Low Accuracy	High Accuracy
Azimuth	183.77	183.76803
Elevation	25.90	25.89698
Refraction	0.03	0.02962
Sunrise	7:19:27 AM	7:19:28 AM
(azimuth)	120.76	120.76363
Sunset	4:35:22 PM	4:35:23 PM
(azimuth)	239.24	239.24052
Transit	11:57:24 AM	11:57:25 AM
(azimuth)	180	180
(elevation)	25.99	25.98909

Figure 5 - The output values for the Ault case

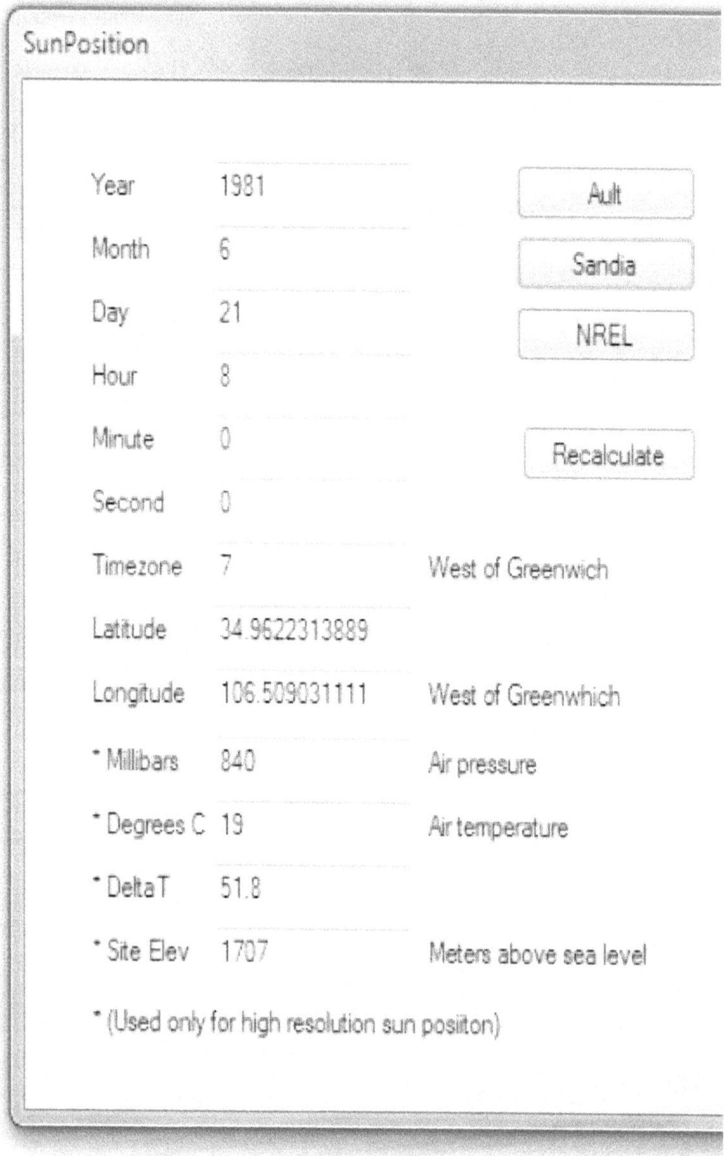

Figure 6 - The input values for the Sandia case

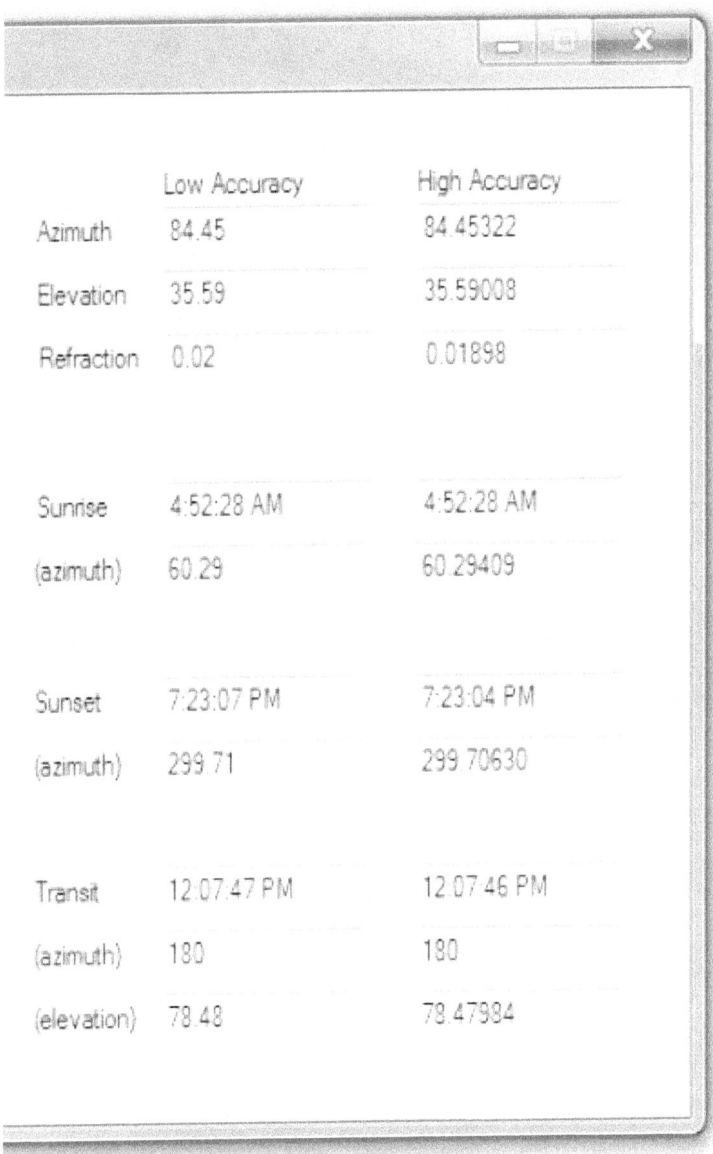

	Low Accuracy	High Accuracy
Azimuth	84.45	84.45322
Elevation	35.59	35.59008
Refraction	0.02	0.01898
Sunrise	4:52:28 AM	4:52:28 AM
(azimuth)	60.29	60.29409
Sunset	7:23:07 PM	7:23:04 PM
(azimuth)	299.71	299.70630
Transit	12:07:47 PM	12:07:46 PM
(azimuth)	180	180
(elevation)	78.48	78.47984

Figure 7 - The output values for the Sandia case

SunPosition

Year	2003	Ault
Month	10	Sandia
Day	17	NREL
Hour	12	
Minute	30	Recalculate
Second	30	
Timezone	7	West of Greenwich
Latitude	39.742476	
Longitude	105.1786	West of Greenwhich
* Millibars	820	Air pressure
* Degrees C	11	Air temperature
* DeltaT	67	
* Site Elev	1830.14	Meters above sea level

* (Used only for high resolution sun posiiton)

Figure 8 - The input values for the NREL case

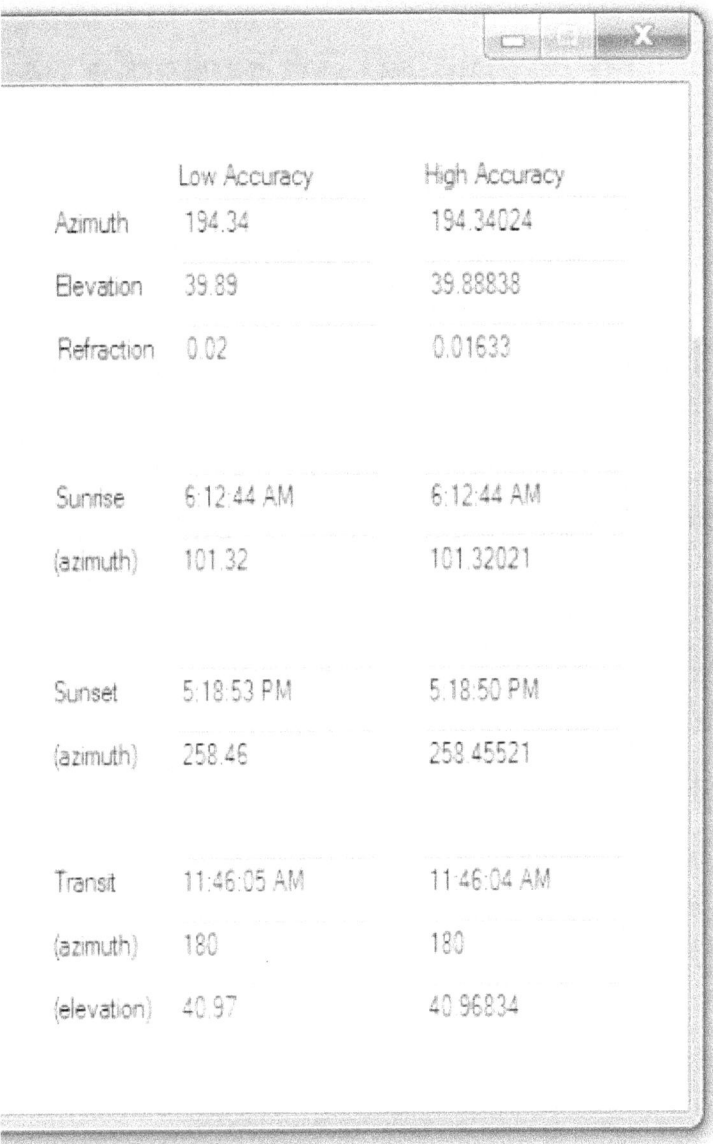

	Low Accuracy	High Accuracy
Azimuth	194.34	194.34024
Elevation	39.89	39.88838
Refraction	0.02	0.01633
Sunrise	6:12:44 AM	6:12:44 AM
(azimuth)	101.32	101.32021
Sunset	5:18:53 PM	5:18:50 PM
(azimuth)	258.46	258.45521
Transit	11:46:05 AM	11:46:04 AM
(azimuth)	180	180
(elevation)	40.97	40.96834

Figure 9 - The output values for the NREL case

Programming By Example
What's next...

Programming by Example by John Clark Craig is a series of books published exclusively for the Kindle platform. They are intended to teach you programming by actually watching a programmer program. Using real-world application program examples, John Clark Craig—author of books published by Microsoft Press and O'Reilly Media— dissects, analyzes and explains the many ways Visual Basic works.

There are many books that teach Visual Basic concepts and theories. This is the first book of its kind to teach programming by programming.

Future books in this series include:

Random Numbers
Working with Date and Time
Graphics
Regular Expressions
Personal Finances
Solar Energy Algorithms
    ...with many more planned after that!

If you have an idea for a book topic or would like to be notified when new titles are released, Mr. Craig would love to hear from you - please write to John@JohnClarkCraig.com.

Some of the other books written or co-authored by John Clark Craig

How to Investigate UFOs using your Smart Phone
Books To Believe In
Secrets to Creating Passive Income
Books To Believe In

VB-Strings   e-Books To Believe In
VB-Random Numbers   e-Books To Believe In
e-Stereograms   e-Books To Believe In
Spirit Mound   e-Books To Believe In

Visual Basic 2005 Cookbook   O'Reilly Media

Microsoft Visual Basic 6.0 Developer's Workshop
Microsoft Press
Microsoft Visual Basic 5.0 Developer's Workshop
Microsoft Press
Microsoft Visual Basic 4.0 Developer's Workshop
Microsoft Press
Microsoft QuickC Programmer's Toolbox
Microsoft Press
Microsoft Visual Basic Workshop: Version 3.0
Microsoft Press

Microsoft QuickBasic Programmer's Toolbox
Microsoft Press

The Microsoft Visual Basic for MS-DOS Workshop
Microsoft Press

100 ready-to-run programs & subroutines for the IBM PC
Tab Books

IBM PC Graphics
Tab Books

True BASIC
Tab Books

101 ready-to-run programs and subroutines for the IBM PC jr
Tab Books

Programs for the Casio handheld computer
Wayne Green Publications

119 Practical Programs for the TRS-80 handheld computer
McGraw-Hill

NOTES

NOTES

NOTES

NOTES

NOTES

NOTES

NOTES

NOTES

NOTES

_____

_____

_____

_____

_____

_____

_____

_____

_____

_____

_____

_____

_____

_____

_____

_____

NOTES

NOTES

NOTES

NOTES

NOTES

NOTES

NOTES

NOTES

NOTES

NOTES

NOTES

NOTES

NOTES

SUN POSITION
IS ALSO AVAILABLE
IN THE KINDLE FORMAT.

IF YOU LIKED SUN POSITION,
PLEASE REVIEW IT ON
AMAZON.COM

IF YOU NEED ANY HELP WITH
THE CODE, PLEASE SEND
YOUR QUESTIONS TO
JOHN@JOHNCLARKCRAIG.COM

CONSULTING
AVAILABLE.